You might need a coat and mittens to read about the winter. You might want earplugs for when they start up the rusty tractor. But an interest in potatoes in is not required—you can get that here. Ash, with a light touch, makes it all fascinating—the farm, the family, the community, and the magnificent, fallible horses.

~ **Sarah Lindsay**, National Book Award finalist, author of *Primate Behavior*

***

Holding the Lines shows that beautiful things can happen when you're open enough to trust your gut, upend your plans, and let life surprise you. Maureen Ash's beautiful reflection on love, farming, and the natural world is a welcome reminder to celebrate hard work, family, and the life you build one foal and potato plant at a time.

~ **Kim Dinan**, author of *The Yellow Envelope*

***

Maureen is a masterful storyteller as she carries the reader along with her in describing intricately and delightfully her fascinating life story. The miracle of birth, the rigors of raising, training, working, and eventually burying an animal that is a partner in a cultural endeavor is only available through living it—or in this case, reading this book. The revealing of details in this family's life is clear, precise, insightful, and shared in a wholesome way in this woman's wonderful literary work.

One of the important "lines" in *Holding the Lines* is the thread of cultural continuity that the horses provide as her Maureen and her family strive to farm in the countryside of Wisconsin. The theme of making a place in their lives for these animals is central to their agricultural approach, and the relationship with the horses is a highly informative part of the story. Keeping work horses in these modern times is often thought to be a romantic, idealistic, non-commercial choice. Yet, there is a dignity available in that choice that is not computed in a financial plan or on a spreadsheet. The sense of comfort that comes from belonging to the natural world is a gift, and humbling oneself to the limitations of an animal-powered enterprise can only be realized by experiencing the life as lived. I found the experience of reading Maureen's book to be both grounding and elevating. I am proud to call Maureen my friend.

~ **Jason Rutledge**, Ridgewind Suffolks, founder of Healing Harvest Forest Foundation

Holding the Lines

Horses, Hard Work, Love, and Potatoes

By Maureen Ash

Acres U.S.A. Publishers

About Acres U.S.A.

Founded in 1971 by Charles Walters, Acres U.S.A. emerged from the need to promote ecological farming practices in a time when industrial agriculture was heavily reliant on synthetic fertilizers and pesticides. Inspired by figures like Rachel Carson and Dr. William Albrecht, Walters used the magazine, and later books and conferences, to advocate for sustainable agriculture that prioritized soil health and natural processes. Acres U.S.A. provided a platform for these ideas and helped to popularize alternative methods like cover cropping and integrated livestock management.

Though the agricultural landscape still relies heavily on conventional methods, Acres U.S.A. has been instrumental in the growing movement towards regenerative agriculture. By disseminating knowledge and supporting eco-conscious farmers, the company continues to champion sustainable practices through its publications, conferences, and online resources, contributing to a shift towards a more environmentally sound approach to farming.

Find Out More About Acres U.S.A.

Subscribe To the Online Magazine
(https://members.acresusa.com/)
Attend The Eco-Ag Event
(https://www.acresusa.com/events/)
Visit The Acres U.S.A. Bookstore
(https://bookstore.acresusa.com/)
Join The Free Newsletter
(https://mailchi.mp/acresusa/newsletters)

Holding the Lines
Copyright © 2023 by Maureen Ash

All rights reserved. No part of this book may be used or reproduced without written permission except in cases of brief quotations embodied in articles and books.

The information in this book is true and complete to the best of our knowledge. All recommendations are made without guarantee on the part of the author and Acres U.S.A. The author and publisher disclaim any liability in connection with the use or misuse of this information.

The author and publisher expressly prohibits any person or entity from using or reproducing this publication to train artificial intelligence technologies or systems to generate text, including, without limitation, technologies or systems capable of generating works in the same style or genre as this publication. The author and publisher reserve all rights to license uses of this publication for generative AI training and development of machine learning language models.

Cover photography credits: Rob Meyer photo

Acres U.S.A. LLC
PO Box 351
Viroqua, Wisconsin 54665 U.S.A.
1 (970) 392-4464
info@acresusa.com • www.acresusa.com

Printed in the United States of America

Holding the Lines / Maureen Ash. Viroqua, WI, ACRES U.S.A., 2023. 250 pp.
ISBN 978-1-60173-1821 (paperback)

Library of Congress Cataloging-in-Publication Data
Name: Ash, Maureen 1957 – author
Title: Holding the Lines: Horses, Hard Work, Love and Potatoes
Description: Viroqua, Wisconsin: AcresUSA, 2023
Library of Congress Control Number (LCCN): 2023948107
Memoir. 2. Agriculture.

ADDENDUM TO HOLDING THE LINES

This book was published in 2023, and some product names mentioned in the book have changed since then.

One product, PhytoMast®, is the officially registered name for what was called Phyto-Mast in this book. The product is an herbal blend of thymol (antiseptic) balanced with angelica, licorice, wintergreen, olive oil, and vitamins A, D, and E. It is packaged in aseptic infusion tubes (with alcohol pads) to be used as intramammary infusion for healthy cows at dry-off. Peer-reviewed, published articles in studies done with dairy cows and goats have shown its utility and safety as a dry-off product. PhytoMast® is labelled specifically with commercial Grade A inspected dairy farms in mind so the farm can meet the regulations of the Pasteurized Milk Ordinance (PMO) item 15r (Storage of Drugs and Chemicals).

Along the way, I enhanced the formula and added cinnamon bark oil (CBO) due to many published studies in the literature demonstrating the destructive effects of CBO upon biofilms that various pathogenic germs form to protect themselves. To keep these similar yet different products separate, the name UdderWell is used for the formula with CBO. UdderWell should not be on the shelf of Grade A inspected dairy farms. In other words, UdderWell is labelled for those who are raising animals for themselves and not for commercial use. The difference between the two products, other than the presence of CBO, is strictly a labeling issue and not a quality or concentration difference.

UdderWell, GetWell, BreatheWell, EatWell, LivWell, BreedWell, and FeelWell are available at www.reverencefarms.com

Pages	New product names
331	**UdderWell/ PhytoMast®** for step (3)
206, 220, 238, 243–245, 269, 290, 373	**GetWell** for herbal antibiotic tincture; for herbal antibiotic mix; for Phytobiotic; for tincture of garlic, echinacea, goldenseal, wild indigo, and barberry
210, 218, 221, 223, 225, 233, 373	**EatWell** for tincture of ginger, gentian, nux vomica, and fennel; for Digestive Ø; for Phyto-Gest
210, 214, 217, 373	**LivWell** for tincture of red root, celandine, milk thistle, dandelion, and oregon grape root; for Phytonic
373	**FeelWell** for Phyto-gesic; for herbal pain formula tincture
284, 289–292, 373	**BreedWell** tincture / **HeatSeek** powder for "Spectra 305"
232, 238, 239, 243–245, 302, 327	**AmpliMune®** for ImmunoBoost; see www.novavive.ca
232, 238, 244, 245	**BoviSera®** for Bo-Bac-2X/ Quatracon/ BoviSera; see https://colorado-serum-com.3dcartstores.com

Dedication

For Richard, Marian, and Truman.
And for the horses, the cats, and the land.

Coins for the Journey

On a hot day the horses find shade
where a box elder has reached some limbs
over the fence

and they stand, drowsing and switching
their tails, lifting a hoof and setting it down
to discourage the flies

or they walk, unhurried, to the pond
and watch the old gelding lie down in the water,
groaning his pleasure

and they let the wind switch its light tail
around their bodies. Small birds come
to sit on their broad, hot backs

as the red, light-edged horses graze their way
back to the shade, another summer afternoon
sewn into the lining of my coat.

MAUREEN ASH

Prologue

It's a relief to turn onto our long, steep driveway after being away. The car tunnels through the shade as it dips down into the section of road my then-small children called Fairyville. It rises into sunlight as our house appears. I'm glad to be home, though I had a lovely time over the weekend.

I ease the car into the shade of the garage and sit for a moment, still buzzing from the road, and I see an odd shape on the lawn. I get out and investigate. In the ell made by the daylilies that rim that side of the house, what looked from the car like a dead Labrador-sized dog is a dead foal—so newborn that it is still damp, lying flat on its skinny side.

No one is around. The truck is gone, meaning my husband is not here. None of the mares were pregnant—that I knew. This foal, however tiny and lifeless, is a Suffolk draft horse. We have one of the few herds of that breed in the Midwest.

This sweet, homely baby is all legs and ribs and perfect details. I kneel to stroke the cold neck and wish her well on the other side, apologizing for it not having worked out over here. But under my hand, she moves. In her little premature way, she says she is not dead yet.

I'd spent the weekend in a small city of leafy, brick streets, sitting up in the evenings outdoors on elegant teak furniture, being treated like a queen by my friend and her two small girls. Those memories evaporate with the road buzz. How did this filly get here, where is her mother, how can we save her? I'm back on the farm with a thud, and the one thing I know is that I have a lot to do, and it can't be done in these silly flower-print town pants. I go inside to change.

A minute later I'm back outside, covering the filly to keep her warm (though the day is hot), and hurrying up to the barn. Still no one in sight. I go to the pasture gate and look out. In the low spot, near the cottonwood tree island, I see a strange pickup truck, a mare, and my daughter, trying to catch the mare.

Then I see our vet, Dick, and I understand their plan. Marian is standing on the tailgate as he drives slowly, carefully, back up to the barn, followed by the mare—it's Jenny I see now—and he stops at the gate. Marian slides off the gate, comes around the truck, and sees me. She bursts into tears and runs into my arms.

"I was riding Cutter and I saw a horse alone, away from the herd, so I went to check, and it was Jenny, and she had twins, and one was dead but one was alive! I carried it up to the house and called Dick. Now we can't catch Jenny, and the other foal will die!" Heaving with sobs, sweaty and tear-streaked, she hangs off my shoulders. But not for long. She knows we have to catch Jenny to save the other foal.

I get a bucket of oats—that most useful item in the horse-catching toolbox. Jenny seems to have given up by then and lets me slip a rope around her neck. Then put a halter on her head. Dick carries the dead foal, which he had lain on the tailgate beside Marian on the ride up to the barn, through the gate, and Jenny lets me lead her through to the harnessing area, where Dick places the foal's body on the feedbox, right in front of Jenny's nose.

"I'll have to milk her for her colostrum to give to that filly," he says. "Having this one near her might help with letting it down."

Marian and I stand by Jenny's head, soothing her as Dick milks Jenny's tightly swollen udder. She sways and makes half-hearted attempts to kick, but he is used to dodging the hooves of upset animals and keeps himself safe. He squirts the orange-tinted colostrum into a plastic container until he has two or three cups of it. Then Marian and I carry the living filly from the lawn up to be near her mother, and Dick snakes a tube down her nose and pours an initial feeding into the sweet little foal.

"After this," he says, "Use a bottle. She should start to suck."

He starts packing up his gear, but something was clearly on his mind. Finally he turns to us, where we kneel by the foal, stroking her body and cooing nothings over her.

"She looks pretty good on the outside," he says, warning us. "But in a lot of these cases, the kidneys haven't developed yet, and they can't process the milk. I hope I'm wrong."

I stand and help him carry his equipment to the pickup and open the gate and close it after him. Dick also has an animal-loving girl at home. "Just keep her fed and warm, I guess. It's worth a try. Good luck." We look at each other, in that moment more friends, neighbors, and parents than vet and client.

"Thanks," I say. "Thanks for coming when she needed you. Thanks for everything." Then I go to look for a bottle and nipple.

Richard comes home from where he's been, and Truman comes home from working at the bike shop. Richard and I puzzle out how Jenny could have got pregnant. There had been a stallion who got into the mares-we-don't-want-bred pasture and spent some time there before I realized what had happened. I'd called Dick to give shots of progesterone to prevent a pregnancy from occurring, but, as Dick explained later, that doesn't always work.

Truman does what he's done all his life—joins Marian in her current adventure. The two of them make a bed for the foal—Marian has named her Sweet Pea—and set up a little camp for themselves. They'll spend the night out there, waking to feed Sweet Pea every couple of hours and making sure she is warm.

Richard and I leave them to their project. This started long ago—I'm remembering a time when we'd kept horses down the road from where we lived before moving to this farm. I'd gone to feed them and found a litter of kittens scattered across the barn floor. The mother must have been very young and surprised by what was happening to her. She was taking care of a couple of them, so when I put the others with her, she accepted them. But one of them, a little straw-colored, mouse-sized kitten, was dead. Or so I thought. I picked it up to toss into the woods beside the barn, where maybe an owl or something might eat it, and I realized that though it was cold, it was alive—barely.

I was afraid I'd crush it if I put it in one of my pockets, so I tucked it into my cleavage and let my bra hold it up. I zipped my jacket up to my neck and hoped I was putting out a lot of heat. I did the chores and returned home.

Marian, barely three years old, was waiting for me on the stairway. "Look," I said, unzipping my jacket and rummaging down into my sweatshirt. I pulled out the damp, limp kitten and held it in my hands in front of her. "We have to save it."

Her eyes met mine, wide and uncertain. I watched them change—she rose to the occasion. She would save it.

And we did. Warm towels, warm water with a bit of milk, a brief spell in the warm oven—that kitten began to stir and mew. We went together to bring it back to its mother before it could take on the smell of our house, and we watched gladly as the cat licked it and snuggled it into the litter.

"Oh, dat's good," Marian proclaimed. "She is wiv her muvver."

Now, years later but no less determined, my daughter is going to save this filly. I bring food out to them, and as I approach I hear her talking to Sweet Pea, telling her about how someday she'll get up and run. "You'll be a naughty little horse," Marian says, with such love and hope in her voice that I have to stop for a moment to collect myself and wipe my tears before I let the kids see me.

The next morning I wake before the sun is up and go out to check on them. My children sleep, as does Sweet Pea, all of them scattered like a litter—cats, kids, and tiny horse, all of them touching each other in some way. Marian wakes and quietly says, "Sweet Pea put her head up when I fed her!"

My heart lifts. The heartbreak that was coming might not happen after all.

Sweet Pea lifts her head a few times that early morning but then grows too weak. She becomes too weak to suck, and she grows colder. Marian stays with her until she dies—Truman had to go back to work. I dig a hole in the garden and carry Sweet Pea's brother out and place him in it. I don't cover him—I know his sister would want me to wait—they spent nine or ten months curled around each other. Now they can spend forever like that.

We carry Sweet Pea to the garden and lay her next to her brother. It takes both of us to lift her, and I wonder how Marian had the strength to carry her all the way up from that low spot in the pasture—so far, crying and stumbling under the awkward weight.

I say something about how lucky we were to have known her, and how much she'd been loved in her short life—the usual things you say when a baby dies—as if that should ever happen, as if that should ever become something you get used to—and Marian says she wishes she could have known her better. Then I shovel the dirt back into the hole and arrange the squash vines to grow across it through the summer. I hug my beautiful daughter. And then we go on with our day. We both grew up on farms.

Chapter 1

We haven't always lived in this place. It seems hard to believe it now, but at one time we lived in a split-level home on a five-acre lot, the narrowest end of which rimmed a lake, where we swam, canoed, and, I thought, one day would sail in a boat I would build in the outbuilding behind our house.

From the window over the sink where I washed dishes I could see the Herrs' house on one side, and from the doorway I could see the Appels' house on the other. They were nice people, helpful and friendly when we asked questions about the workings of the neighborhood, but they had moved to what they thought of as the country to be left mostly alone. I was asked to be part of Mrs. Appel's bridge club—a kind gesture on her part, but at the time I saw it as the first sign of my becoming a Stepford wife and I declined. Otherwise, we were left to ourselves and expected to do likewise for them.

I found it confusing, having grown up in a rural area where the neighbors valued each other and were keenly interested in each other's concerns—whether they wished to admit it or not. We could see each other coming and going, mowing our lawns, working in our gardens, for Pete's sake. Why this studied ignoring of one another?

Oh, well. I had things to do. Our garden grew steadily bigger. Richard planted a patch of wheat and I loved the look of it—the matte green color, the way the wind stroked it. It turned gold and dried, and my parents, both of them children of wheat farmers, came down and Dad showed us how to use a scythe and cut it, then stand it in shocks to dry.

Richard was satisfied then, but I had to go the next step and thresh it. I used our daughter's plastic swimming pool, dried it out and hit fistfuls of the wheat plants into it, knocking out the wheat berries, then took a window fan and blew off the chaff. I put my hands into the mostly clean wheat berries and cupped them and raised them up and let the wheat dribble down through my fingers like a miser loving her gold. But you can't eat gold coins, and you can't plant them to make more. Wheat berries are wheat seeds. You can grind them into flour, you can plant them in the ground. They're always cool to the touch, it seems to me, and I have not met many people who, upon seeing wheat in a bucket or bowl or gravity box, have not done what I did—held it in their hands, felt rich.

Richard bought a small, cast-iron mill for me, and I ground the wheat into a coarse flour and made a low, rectangular, dense loaf of bread. Richard's older girls, who lived with us then, tried to be polite about it but could not help laughing when I tried to cut it and kept repositioning the knife as it slid off the loaf's hard surface.

"Make a few more of these," suggested Richard. "We can build that brick wall you've been wanting."

Well, okay. It was not a success. But I kept the mill. I cut a few stalks of unharvested wheat and put them in a vase and liked how they looked.

We lived more and more in the garden. Marian toddled among tomato plants and beans and in the fall waited like a small vulture beside me as I dug the potatoes. She loved finding them in the dirt, piling them in the bucket, riding on them heaped in the wheelbarrow as I trundled them back to the shed.

On an April day I was standing beside the sliding door to the deck of our bedroom and watching it rain. Richard came in and said, after a pause, "I really want to farm." He was an environmental toxicologist, a geek with a PhD in biochemistry. "Never marry a farmer," my mother used to say in half-joking exasperation when I was growing up on the farm and it rained on twenty acres of hay that had been perfectly cured just that morning, or milk prices dropped, or the young stock got out and trampled the garden. Of all the advice she offered my sister and me, that was the nugget we took to heart. We wanted lives without dirt, mud, hay chaff, and sweat. I wanted a life in which weather was just the weather, not the monitor of our fortune, small as that might be. I went to college, but in making that move I was also, in my head, leaving the farm.

And now Richard stood quietly, waiting. Outside the rain-striped window the unplanted garden soaked up moisture. All I remember saying was, "Okay." Maybe I thought he'd get over it. Maybe I thought he'd just rent some land for a season and get it out of his system. Maybe I wanted to farm, too, and just hadn't figured it out yet.

The very next weekend Richard headed off to an auction to buy a tractor. I was at home, it was raining again, Marian was fussy, and the two older girls, Sarah and Emily, were busy in their rooms with their own projects. He'd been at work all week and now he was gone again. It was a long day and I had time to reconsider. Farming! What had I been thinking!

It was a long day at the auction, too. Tractors cost more than he'd expected. Richard came home in a thoughtful mood, and we put his idea on hold for a while.

We were talking about it one evening when I said, wistfully, knowing it couldn't happen, "Too bad we can't just farm with horses." I'd got my first horse, a spirited pony named Splash, when I was nine. I'd begged for two years and finally my dad said that if I could earn twenty dollars on my own, I could get a horse.

My mom helped me find a money-making venture—I sold greeting cards door-to-door. In the country. The real country, not a rural subdivision. I walked miles with my bag of samples, and though I am sure they didn't much need my all-occasion, birthday, and sympathy cards, the neighbors kindly let me write down their names. I rode my bike or sometimes Mom drove me when the orders came in and I had to deliver the goods and take the money. One happy day I had twenty dollars to hand over to my dad, and we headed that very Sunday to the auction.

I had horses after that, through high school. Dad sold my mare, Keko, when I went off to college—we both knew I wouldn't be coming back to the farm. I went four years without riding a horse, and then—after graduating early, taking three jobs, and living pretty much off saltine crackers and a daily orange—I had enough money saved to go to Europe. My friend Beth and I hitchhiked through the summer and then I found a job at a stable in Ireland, exercising steeplechase racers.

"You can ride?" my prospective boss asked me.

I hadn't for a long time, and never in an English saddle—rarely in a saddle at all—but I said yes. I got the job and rode several horses a day and then, on my half-day off, rode at a nearby stable—supposedly lessons, but mainly we arrived to already-saddled horses, swung on, and followed a course through the countryside at a gallop, jumping anything that got in our way.

It was a dream job, but eventually I had to go home and start my life as a grown-up. Now, grown up, I thought maybe a farm where horses were the power source would be a farm where I'd feel more like staying.

Richard cocked his head, thinking. I knew that look. He wasn't ignoring me; he was calculating, evaluating, pondering. "We could do that," he said. "We want to be sustainable, and what could be more sustainable than using horses to farm?"

I don't think the answer to that question is as obvious as we thought it was then. But with that significant revision, the noisy, loud, machine part of farming fell away, to be replaced by romantic scenes of draft horses with me holding the lines. In a way I'd never imagined before, I could see myself being part of Richard's dream. It became mine as well, and we were now going to farm.

***

I met Richard when I was teaching at a junior high school in Minnesota, and I was teaching because I'd been a magazine editor in Illinois, where from my office I could look into the window of a house across the street and lock eyes with a cat, who sat in that window and looked out at the world and had private thoughts about it.

Something about the way that cat sat in the window, just as I stood in mine, made me think there had to be a better way to spend my life. I left that job and returned to Minnesota, my home state, and got the teaching job. I learned quickly that I did not like teaching, but the kids were fun and there was an old swimming pool in the basement of the school where I showed up every morning, with a gaggle of others, to swim laps.

I ran my first marathon that winter, the Torturous 26, part of the St. Paul Winter Carnival. In spring I ran a fifteen-mile race from my hometown to the neighboring town. One of my swimming buddies and his wife and kids came along to cheer me on. Monday morning in the locker room he told the other guys about it.

At the pool that morning, the man in front of me said, "I heard you ran a race this weekend." I could barely make him out through my swim goggles. I was at the part of my workout where I did wind sprints on the minute. I had only a few seconds to say, "Yeah, I did," and then I was off.

He was waiting for me when I sprinted back. "A fifteen miler?"

"Uh huh." I told him my time—nothing to brag about. I'm slow. Big deal. And then I was off again.

This is how our first conversations took place. Richard wasn't a regular. He was there to practice for a triathlon. He swam laps until he saw me pause at the edge of the pool, gearing up for my wind sprints. Then he made the best of his five or fewer seconds as I panted, resting, before I took off again.

When I was in junior high school, I saw my friends undergo a transformation. Whereas they had been funny and smart and interested in school, now BOYS were the thing. They changed who they were, tilted their heads in artificial ways, started wearing makeup.

I had five brothers. Nothing about me wanted to bring another male into my life unless he would be interested in me at least as much as I was interested in him. At that time, the train tracks across the highway from our house were still used by freight trains. It was easy for me to envision how I wanted to be. I thought, 'I'm going to be a train. Nothing will get me off my track. If a boy wants me to like him, he'll have to get on my train.'

More than a decade later, I was still pretty much operating on the same principle. I had wind sprints to do, I had a limited time during which to swim my 1.7 miles, and I was not going to stop and flirt. I knew this man was interested in me, and he seemed like a nice guy. Too thin for my tastes, and I was never one for beards. On top of those drawbacks, I'd learned that he had three children from his previous marriage, and two of them lived with him. I was twenty-seven years old and perfectly happy with my life as it was (except for the teaching). It was no effort at all to keep swimming.

One day I came out of the locker room into the street and the man was standing in the rain. The hood of his jacket was up, but it couldn't protect the water from beading on his glasses and dripping from his beard. I could see, now that I was not wearing foggy swim goggles, the kindness in his face.

"I was wondering if you'd be interested in a picnic," he said.

We made the arrangements, and I went up to my classroom. Huh, I thought.

Over the next eighteen months, after which time we married and moved out of that neighborhood, we ran a lot on the little roads around his house. There was one farm we loved passing because a few Shire draft horses grazed in the pastures. We told each other we'd farm someday and use draft horses. Well, why not? Everything seems possible when you are in love. Though that particular goal did not seem very imminent.

***

It's one thing to decide to farm, however, you wish to farm, and it's quite another to actually do it. And between those two things, as you can imagine, there is time. And during that time, a lot of thinking and imagining and working things out. Discussion. If this were a film, we'd cut now to shots of Maureen and Richard reading from a stack of magazines called The Small Farmer's Journal, attending horse-powered farming events put on by enthusiasts of old-time farming methods in other parts of the state, subscribing to *The Draft Horse Journal*, visiting a local horse trainer, and talking. Talking and talking.

"I have an idea," Richard said. "What if we don't just farm with horses, but we breed them, too? We could choose a breed we like, buy registered stock, and then raise and train the foals. The more we use them, the more valuable they are—it's the opposite of a tractor! Then we sell the trained teams. It's another income stream."

"We don't even know how to harness a horse yet, let alone train one," I reminded him. But this was not an unusual exchange—as you will see.

The original plan was to rent some land in the area. It wasn't that we liked our house so much—it was a shoddily built split-level from the 1970s. It made poor use of heating and cooling dollars and was unattractive to boot. The location was great, though—on a small lake, near a county park, next to a nice bike trail, in a good school district where the older girls were reasonably happy, not far from Richard's job.

He would keep his job. I'd keep my little writing and editing business. I hadn't lost my mind! I wasn't going to go back to living as I had on the farm growing up. We'd farm, yes, but not depend on it for our livelihood. The romance of farming with horses couldn't begin to make up for the reality of what I remembered from my youth.

Those were still horse-and-buggy days as far as gathering information was concerned. We wrote to people we read about, and quite often they wrote back. We called people on the phone—our silly phone with its curly cord hooked into the wall. We decided to see for ourselves some of these horse breeds we were reading about. We arranged for our summer vacation, when the older girls would be with their mother, to be a cross-country trip to visit Richard's parents in northern California. We drew our route dot-to-dot, each dot representing a different breed of horse.

It was in Alberta, Canada, that we saw our first Suffolk draft mare. Her name was Marigold. She fit my eye. She was the size and shape we'd been looking for. The look she gave me was mild and intelligent. I loved her chestnut color and the neatness of her legs.

Richard and I agreed. Suffolk. Suffolk horses. We'd farm with them, breed them, raise and train them. Using them for power would add to their value, and we'd sell trained teams. It would be an income stream.

Well, that was figured out!

Back at home, we carried on as usual. Work. Garden. Run. Look for fields to rent. Figure out where we'd keep our team. Find a team.

Right. Turns out, it's not that easy to find a team of Suffolk draft horses that are trained enough for beginners to use around their little girl. Also, there was still that bit about our not knowing the first thing about harnessing, driving, and working horses.

In fixing the latter, we also solved the former.

***

"Look," I said, lifting our cat out of the way and making room for Richard on the couch beside me. I was reading The Small Farmer's Journal again—someone had given us a stack of them, and with each one being oversized, as well as fairly thick, we'd be reading through them for some time to come.

It was an ad for a five-day course in learning to harness, drive, and work horses, being offered by a couple who bred Suffolk horses! Wow, we thought. Sounds perfect.

Ron met us at the airport in Knoxville. I'd been nervous about this trip—were we wasting our money? How could we learn enough in five days to get ourselves started on this ridiculous adventure? But, shaking hands with Ron, I relaxed. Early thirties, easy going. "Hey," he said. "Good to finally meet you."

His wife, Deb, was waiting at the farm. She was small and rail thin, but there was no hiding the strength emanating from her. I felt her size us up, and I hoped she approved.

"So why do you want to farm?" she asked us over dinner. I waited for Richard to answer. It had been his idea, after all. This seems very odd to me now, but at the time I just accepted that he wanted to. And why wouldn't I? In my experience, men wanted to farm. All my grandfathers on both sides, as far back as I knew, had been farmers. My dad was a farmer, two of my brothers were farmers. It was the default setting for a guy, I thought.

Richard cleared his throat. I felt him reaching around inside for his words. "I just always wanted to," he said. "My parents had a little hobby farm and I raised pigs when I was in high school. We had a donkey. I was in Future Farmers. I guess I really did think back then that I'd farm."

Ron leaned forward. "So what got in the way?"

"Oh, well. California, I guess. That's where I grew up and thought I'd always live. And, even then, the land was so expensive. I figured I could never afford to farm, so I went to college."

I laughed. "I know that's true," I said. "But it just sounds so funny to me. I think in my family, people farmed because they couldn't afford college."

"I know what you mean," Deb agreed. "Now it's kind of a luxury to even think about farming. We could never do what we're doing if this weren't my family's land."

"It's an interesting question," Ron said, "when you put horses into the equation. Can using horses reduce the cost of farming and make it something more people can afford to do?"

I had to think about that. There was a picture of my dad filling a corn planter out in front of our barn in the spring of 1947. His team, Dick and Dolly, stood patiently—anticipating, no doubt, a very hard day ahead. The caption on the photo is in my mother's handwriting: "Our first spring planting."

My dad was an impatient man, and I'm sure he drove his horses hard. He bought a tractor by next spring's planting. He said himself, years later, when I asked him whether it was better for horses now or "in the olden days, before tractors," that a lot of horses were mistreated and overworked when they were the main source of power.

I told my dinner companions that. Deb was quick to pick up on one thing, though. "But your dad had to wait to buy that tractor, didn't he. Using the horses, even for just the one season, helped him keep expenses down and get the loan from the bank."

"How many banks would loan money to a farmer now who wanted to use horses?" I asked. That was a rhetorical question, of course. Later I would learn there really is an answer.

On into the evening we talked about horses, farming, the state of the world, our hopes. We had found friends.

In the morning Ron clipped a set of lines to the bars of an unused farrowing crate. Inwardly I itched. I wanted to be with the horses!

He showed us how he liked to hold the lines and explained the advantages of that style. "Line management," he said, winking. I could tell he knew what I was thinking! I had to laugh and relax—he'd get to the horses sooner or later.

He showed us how to loosen and take up line and made us do it over and over. Only then did we move on to the harness.

A horse stood tied outside the harness room. She had that same solid, patient look I'd seen in Marigold up in Alberta. I was still hoping we could find a team of Suffolks.

Ron showed us where the brushes were kept and explained how important it was to start with a clean horse under the harness, or the chafing could become a problem.

"Especially on the shoulders," he said. "You watch those like a hawk. You'll take off the collar and see some wrinkles—just like we get blisters, so do they. So fitting the collar is really important."

Pads, collars, different types of harness—I hadn't realized it was so complicated. One thing I had thought to wonder about was how long it would take to harness and unharness a horse. In the pictures I'd seen, the harness gleamed with buckles. Did the teamster have to buckle and unbuckle each one of those?

No. I learned that from Ron that morning. He showed us how to put the collar on the horse and then turned to the harness.

"When you look at this thing," he said, "you think, wow, that's a lot of fastening. But really, most of the buckles are for sizing it. Once you have the harness sized to your horse, it's pretty much just five or six snaps and buckles, and boom—you're done."

He put his arm and shoulder through what looked like a nest of leather straps, grabbed what I would soon learn was a hame, and lifted it off the rack. The horse, Greta, stood patiently as he slung the mess over her back. "Atta girl, Greta," he soothed her. She swiveled one ear back toward his voice. She was used to this process, but still and all, it's nice to be acknowledged.

He settled the metal hames into the crease in the collar that is made for them, and then buckled them snugly. "From here," he said, "everything just kind of settles into place." And it kind of did. We watched him work the straps away from the collar and down Greta's back.

"This is the britchen strap," he said. "You say it 'britchen,' but it's spelled b-r-e-e-c-h-i-n-g."

"Like how 'britches' is supposed to be spelled b-r-e-e-c-h-e-s," I said.

"Well, moving on," Ron said significantly, getting our attention back on the harness itself and not the spelling of its components, "you snap the yoke strap, flip up the pole strap, buckle the belly band, snap the quarter straps to the pole strap, and you're ready for the bridle."

He moved smoothly down Greta's thick body as he spoke, buckling this, snapping that—bing-bang, done. Greta stood harnessed. We learned to put on her bridle and attached the lines. Next step—or so I thought—driving!

"And it's just as easy to take the harness off," Ron said. He proceeded to do so, putting everything back on the rack and standing, arms folded. "Now you harness her."

It's always easier to watch someone do something. It makes so much sense as they narrate their actions. You think you've learned. But then you try it.

Well, at any rate, we got Greta harnessed again, and then we had to unharness her. And then harness her again. And then we began our driving lessons.

How strange it seemed to be so far from the horse. I'd been riding since I was nine, always bareback. My horse and I could feel each other think that way, as I told anyone who asked why I didn't use the western saddle we had in the garage. Well, it was heavy, too. But mainly, I liked the close contact of my skin against my horse's hide.

Driving, I had no contact other than through the bit in Greta's mouth. I could talk to her, which Ron encouraged me to do, or I could flip the lines against her rump, which he did not. How to make her step forward? How to control her turns so she didn't just carry on in a circle rather than the gentle left or right I wanted her to make?

Ron set up cones for us to practice driving around. We stopped, backed, turned. Greta was patient with us, and each time it was not our turn, we watched the other and thought we could do better than that.

After lunch, we harnessed Garnet as well so the two of us could practice simultaneously. We drove our horses into the pasture. I mentally set up an obstacle course—drive to that bush, circle it, drive to that hillock, turn left, drive to that tree, stop, back up, start, circle it, reverse, come back. I was surprised at how well the horse responded to my voice. I was beginning to learn the feel I needed to have in my hands—just the right contact, not too much, not too little. If I increased contact on the left to turn the horse left, I needed to let out line on the right—but not too much. I needed to be able to take it in quickly. Ron's lessons at the pig-farrowing crate began to make sense. Line management.

We progressed from ground driving to hitching the horse to a cart and driving from the little seat above the wheels.

"I like this," I told Ron as I drove Garnet down a little road on the farm. "I'm worried, though, about driving a team. Is it twice as hard?"

"It's easier," he said. "You'll see. Horses are herd animals. They'd rather go with someone else."

He was right about that. Deb taught us how to figure out which is the "long line" to use in a set of lines for driving a team. (Tip—it isn't the longest line.) We did our obstacle courses, ground-driving Greta and Garnet as one team and Jane and Aster as the other. Aster kept hearing her foal back in the barn and wanting to turn that way, so I had to work harder to keep her in line. It was good experience, frustrating as I found it at the time.

We learned to hitch a team to an implement in the safest way. Even today, I follow the steps I learned from Ron and Deb. Their safety routines were what we learned before we could develop bad habits, and I feel it is one of the main reasons we have not, so far, had a bad accident with the horses.

We learned to drive teams of three and four horses. With four horses hitched abreast, line management became the name of the game. The horse on the inside of the turn needed the line taken in and in and in. The horse on the outside needed plenty of line to be able to make the turn. Then it all had to be equalized quickly to keep them on track once the turn was completed.

Ron and Deb had taught themselves all this through trial and error. In five days, we gobbled up what they'd learned—about driving horses, farming with horses, making a living on a mixed-power farm, marketing produce. We were outside with the horses all day, and in the evenings, over dinner and after dinner, we talked.

One of the things we talked about was their mare, Jane. I'd taken a liking to her and suggested to Richard that we try to buy her. And maybe Greta or Garnet to go with her, to form the team of broke Suffolks we had been trying to find!

Ron dropped us at the airport in Knoxville with a promise to consider our request to purchase Jane. I hugged him good-bye with real regret that we did not live in the same neighborhood. We'd come to enjoy their company so much. Not to mention that we could have used the mentoring.

But back to our regular lives we went. We went to pick up Marian at her sitter's house. Through the window we could see her playing with Cindy's, the sitter's, older kids, Miranda and Zach. She was happy. But Cindy had seen us coming up the walk and directed Marian to look out the window. Her face when she saw us—that in itself is almost reason enough to have kids. No one in your previous life has ever been that happy to see you!

Except, I guess—now knowing it from this side—your own mom.

Chapter 2

One of the contacts Richard made regarding land came through. A farmer could rent him ten acres not far from our house. That was enough for a start, we decided. It was becoming autumn. By spring, we felt we could have a team and some equipment ready for use. We could fence in part of our lot and keep our horses there. We'd build a small shelter for them. We'd buy hay for this winter but hope to make some for next.

It was exciting and interesting to me. We kept reading and sharing what we'd learned and going to horse events, and Richard began attending auctions to see what was available regarding horse-farming equipment and how much it tended to cost.

Nothing had held my attention so closely before. I was tired, though, and my period was late. Almost out of duty rather than in any kind of expectation I brought a urine sample to the clinic (this was in the olden days, remember) and found out I was pregnant. Due in May. Just when I'd expected to be plowing with our team, planting our crop—whatever we might decide that would be.

Richard came home from work that day and found me in our bedroom, sitting on the floor and leaning against the bed. "Come and sit down," I said.

I'm quite a bit older now, as I write this, and I have the benefit of hindsight. I do remember feeling as if our dream was squashed—or at least my idea of how I'd be participating in it. I was unhappy. I'd taken precautions. This wasn't supposed to happen. The world is just not fair.

Richard was less disappointed. Typically, he was philosophical. Not delighted—but he could roll with this.

From my current vantage point, I feel sad for us, that we had no idea at the time that this mistake we'd made—somehow getting pregnant—was one of the best things that would ever happen to us. That our son Truman would be such a source of joy and pride and laughter, that he would take us places we couldn't have expected we'd ever go. I'm sorry for him, sweet boy, that his first stirrings were not greeted with happiness.

In our defense, I will say that once we adjusted to the facts on the ground, we did become happy. We told Marian about the new baby in

Mom's tummy and she—generous soul that she was then, and is now—accepted him immediately as "ours baby."

A great memory I have of my dad is of the day he learned I was pregnant with Truman. I'd taken Marian up to see her grandparents, and we were having dinner with them. My youngest brother was eating with us, seated right beside me. He and his wife had recently learned, also to their surprise, that she was expecting. They were rolling with it, too.

"Marian," I said to her. She was sitting by Grandpa, as she loved being near him. She loved her Grandpa to the point that for a period during that winter she begged most often for me to draw a picture of Grandpa for her—which I did, portraying him as an oval with a kind of "greater than" sign for a nose, a straight line for a mouth, rudimentary ears, and some squiggles that were wisps of hair (he was quite bald). Even now, I could draw that picture in my sleep.

"Marian." She looked at me, her face smeared with mashed potatoes. I said, "What are we getting at our house?"

She looked puzzled. My parents looked interested. New car, maybe? I could see them running scenarios in their heads. Not a new TV; they don't watch much of that. Could be a little boat . . . "What are we getting at our house?" I asked again. She thought hard, and then it came to her.

She looked up at her beloved grandpa. "A baby! We getting a baby!"

The shock was not as significant as if I'd said we were getting an elephant, but it was definitely a shock to them. No one had expected another baby from Rich and Maureen. Still, no one was more surprised than Rich and Maureen. I explained that, and my brother beside me commiserated. Dad looked at us, realizing we had, even with the benefit of modern birth control, STILL GOT PREGNANT. Not that we, especially my brother—my youngest brother, remember—minded having been born, but it's true, we'd always kind of treated Mom and Dad's large family as a kind of failing. If only they'd had modern birth control in the olden days—you know, that kind of thing.

But here we were, two of his college-educated kids, somewhat abashedly owning up to having made a couple babies by mistake. And he tipped back his head and laughed at us with such delight, I think of it now with a bit of a tear in my eye. Glad I could make you happy, Dad! As Truman did when he came along, as did his cousin Roy, Scott's unplanned-for (but joyfully welcomed) son.

**

Through the Suffolk Association we learned about a farmer in Minnesota who raised Suffolk horses. In fact, his dad had raised them for years as well. Richard called him and made arrangements for us to come and look at the herd and possibly buy one or two.

Real farmers have a look—ruddy, worn, usually pale of forehead if you catch them without a cap. Jim was definitely a farmer. He held out his big hand to each of us, including Marian, who was in my arms, waking up from the nap she'd had on the drive over, and led us to the pasture.

His horses were beautiful. "You can see the difference in the lines in the older mares. My dad had two stallions, and they each threw a different type. He liked them both, and it's been interesting to see what happens when you cross the daughters on one side to the stud on the other."

I could see what he meant—the Ransom daughters had the finer heads, while the Starfire daughters' heads were, while not coarse, certainly less chiseled.

"I just don't use them anymore," Jim said, regretfully. "I love 'em. But I'm busy with the cows and everything else. The kids are in activities at school and my wife works long hours and it's just the lowest priority, so it doesn't get done."

I don't know what got into us. Well, I guess I do—she was just so beautiful. We picked out a six-year-old mare named Dinah. She was bred to foal in the spring. And she'd never even had a halter on.

Jim agreed to deliver her after we had a fence and shelter built. And to put a halter on her. Finally.

One evening later that fall the phone rang. I was delighted to hear from Ron. He and Deb were willing to sell Jane, and with her they'd also sell a younger, green-broke mare named Jemima. He'd deliver them in the new year. He was surprised and happy to learn that we were expecting another baby. And guess what? Deb was pregnant, too! Due in August. It was a happy phone call all around.

Our fence-making and shelter-building was already in overdrive. One of the things I was learning about Richard through this new farming project was how many things he either knew how to do or could learn how to do. He framed up the three-sided shelter and taught me to nail off the plywood as he moved to working on the roof. Once the plywood was on the roof, we schlepped packages of shingles up the ladder, and he showed me how to shingle a roof. I worked on it during Marian's nap times when he was at work, propping the intercom for the baby monitor against the pile of shingles. It wasn't a big building, so it didn't take long to finish, but I felt proud when it was done. We sided and painted it to match the house. The little paddock was ready—the strands of electric fence ribbon tucked into insulators on the t-posts, the solar-powered fencer neatly stowed on a little pedestal.

Richard built harness racks into our little outbuilding behind the garage. We walked from our house to the rented field, figuring out how to get there on the state trail so we wouldn't have to drive the team on a road.

We could hear the truck slowing to make the turns—first from the county highway onto the smaller road that led to our even-smaller road. It was an auditory lesson on driving with live animals in the trailer—slow, steady, careful. By the time the truck was in our yard, we were dressed and outside, our breath fogging around us. It was early morning. They'd driven all night.

Ron's hug was energetic. "We're all coffeed up!" he exclaimed. "Excited to be here!" He looked around approvingly at our suburban horse paddock. He introduced his friend, Frank, and asked me where Marian was. "Still sleeping," I said. "But you'll see her soon."

The trailer—a gooseneck, which is a type of trailer that connects to the truck via a ball hitch sunk into the bed of the truck—rocked. We could hear the horses shifting within. I was shaking with excitement and cold, ready to see Jane again and to meet Jemima. I must have seen her when we were at Ron and Deb's farm, but they had a lot of horses. She had not stood out.

Jane came off first. Ron gave me her lead rope and I buried my head in her mane, taking in that horse smell I loved so much. Ron went back into the trailer and came out with Jemima. She was slightly taller than Jane, who could best be described as long and low. Jane had a small star on her forehead. Jemima had a narrow blaze. Jane was a dark chestnut, Jemima slightly lighter. She seemed worried in this new place, which was understandable.

We led them to the paddock and then led them around the inside of the fence, acquainting them with their new home. We have since tried to do this with new horses. It isn't always easy, but it has always seemed like the polite thing to do for them. How awful to have one of the established horses in the herd chase you, the new horse, and you only see the electric fence for the first time as you run into it! Showing them the parameters of the pasture can help avoid that.

Then we went in for breakfast. Ron and Frank met Marian, who was delighted to have two new people to inspect, question, and use as furniture.

"So. Not that shy," Ron said dryly as she parked herself across his lap.

We'd also purchased harness from Ron and Deb. Deb had her own harness shop—one of the ways they kept afloat—and she'd sent three sets up with the horses. Ron helped us fit one set to Jane and Jemima, and then we harnessed them and drove them around the yard, down the driveway, and back and forth on our little rural street, walking behind them and talking.

"You can see Jemima's worrying," Ron pointed out. It was true, Jemima's ears were active, and she was jumpy. "Jane's just going along. The more you drive them together and nothing bad happens, the more Jemima will settle into it. Jane's a great teacher, just by going along the way she does."

Don't let anything bad happen. How many times have I thought and said that while working with horses, advising people? Since Jemima, I have trained dozens of horses, and still that is the central tenet. Horses have to believe you know what you're doing. Nothing bad happening is proof that you do.

We hitched them to the bobsled Richard had bought at one of the auctions he'd attended, standing for hours and taking his careful notes, finally bidding when he felt he could get a good deal. Driving our fluffy, beautiful, new team around our neighborhood felt wonderful. Taking off the harness, rubbing their sweaty spots with wisps of hay, turning them out into their paddock, stowing the harness on the racks we'd prepared—Richard and I sometimes caught each other's eye and knew what the other was thinking. It's happening. It's starting to happen.

It was as if a long-planned journey was now underway. The train was leaving the station.

Chapter 3

Can I reach back from here and just guide our hands a bit? I could be so helpful to us if only that were possible. Alas, we had to make our way, after Ron left, by ourselves. And we were not, at that point, fully capable of ... well, much at all.

I was in the house with Marian, just awakening from her nap. Sarah and Emily were with friends. Richard was outside, practicing with the horses. He'd built a small wooden sled that he could stand on as he drove. It was a terrible idea, but who knew it at the time? Driving from a standing position is for drivers who know what they are doing. Beginners should sit. This sled didn't have a pole, but was pulled by an evener that was clipped to the sled. That meant it could run up on the horses' heels—a great way to spook them.

The door opened. I could hear Richard's heavy breathing. "Maureen," he called, weakly. But I was already on my way. He was seated at the bottom of the steps, hunched over. No blood, just broken glasses and torn clothing.

"What happened?"

"Runaway," he said. "They just ran away. The sled went over me. I'm kind of surprised I'm still kicking."

"Where are they?"

"Tied up. Kind of. I'll go. Back out." But he needed to NOT go back out; I could see that. I was already putting on my winter clothes. Marian came from her room and slid down the stairs to be with her dad, who pulled her onto his lap and held her as would any dad who just recently had contemplated his untimely end.

He was catching his breath and more able to explain things now.

"I was riding on the sled, and they were going fine. We were on the soybean field across the road." (We'd checked with the farmer and had permission to drive on his winter fields.) "Jemima was just feeling good, I guess, and she pulled ahead. And I pulled her back, but too much. So the sled rode up on them, and I lost my balance and fell over the front of the sled and I could feel the backs of their legs and then the evener and the sled going over the top of me and rolling me along and then—they were off. Gone. They ran like a house on fire to the edge of the field and hung up on either side of a tree. The harness snaps and buckles broke. A couple straps I had to cut with my knife to get them out."

He winced as Marian passed her hand wonderingly along the side of his head, where it was rubbed almost raw.

"I left the sled and evener and ground drove them back here. They're tied up. I'll be out in a bit," he said, as I was slipping on my gloves to go out.

"Make sure Marian wears her hat," I said, shutting the door behind me. The cold was useful to me in that moment, near tears as I was. What were we doing? Marian could have lost her dad. The baby inside me would never have known his father.

I found the horses tied to the rail Richard had made for that purpose. Ideally, I'd have taken them back out and driven that old memory away. But the harness was in tatters. I was more worried about Richard than anything at that point. They were sweaty and blowing. I took off the harness that wasn't already dragging, and I rubbed them with handfuls of hay. When they were as dry as I could make them, I put them into their paddock and spread the harness on the floor of the outbuilding to assess the damage.

But I didn't take time to look at it much. I threw hay to the horses, made sure their water was free of ice, and went back in. We were going to have to think about things.

"What do you want to do?" I asked him over dinner. Marian cocked her head interestedly. It made us both laugh.

But he knew what I meant. "I want to farm," he said. "With horses. This was a setback, but we'll have those. This was probably just the first."

I knew he was right. "Let's hope it was the worst," I said. "And that Marian is never involved in anything like this." That was the thing we feared most—that Marian, and maybe someday this new baby we were waiting for, would get hurt because of this dream of ours.

After dinner we went out to the outbuilding and looked at the harness. We figured out what needed to be replaced, and we wrote it down. In the morning, I called Deb's harness shop and made the order—glad, kind of, that Deb hadn't been the one answering the phone and taking the order. I didn't want to have to explain ourselves just yet.

Within the week the straps, snaps, and buckles had arrived. We'd been handling the horses every day since the incident, walking them up and down the road to keep them familiar with their surroundings, with us, and with being out of their pasture. We put the harnesses back together, lifted them onto the horses, and started from the beginning, ground-driving them singly, then together. Up and down, here and there. Day after day after day. We kept the bridles in the house so the horses didn't have to accept a cold piece of metal into their mouths. They were such good mares that they would have, but it seemed unnecessarily cruel.

It took a long time for Jemima to trust us again. Something was gone from her that we just could not get back. Was it her nature, and she'd have been like that no matter what—always just the tiniest bit worried around us? Horses are so much like people. Some just have a natural confidence, and that makes them easy to train. The ones who lack it, even a little, are never fully at ease. On the other hand, the luckier ones of that type sometimes find solace in a sympathetic and confident handler.

***

Dinah arrived; Jim pulled the trailer into the yard of the hobby farm down the road where we'd made arrangements to keep her. I put Marian in the car and shut the door to keep her safely out of the way, and I stood well back from the trailer, expecting Dinah to burst out once the door was open. I pictured poor Jim hanging off the rope on her halter.

Instead, the door opened and Jim led her off. No fireworks, no drama. I showed Jim where she would go in the pasture and we walked her around the fence, punching through the snow, so she'd know her limits. Then we showed her the water and threw her some hay.

"Looks like she's home," Jim commented, looking pleased.

It did look like that. Dinah was content, as she would be for years with us—as long as we didn't try to hitch her. We had no business buying her, green as we were. She was gentle and friendly, and we could harness her without trouble. She exploded into kicks and bucks if anything was hitched to her, though. We took her to a professional horse trainer, who kept her for six weeks and then returned her one rainy day when we were not home. Friends of ours called him a week or so later, pretending to be interested in buying her and wanting his opinion of her. "She's crazy," he said. "I've never seen anything like it. Don't buy her."

We resigned ourselves to keeping her as a brood mare. I know we were taking a chance—we knew it then. Would her nature be passed on to her foals? We decided to proceed on the information we had—her sire and dam had been good workers, well known in the breed. She'd never been handled or haltered till she was six. We did not know how she'd been treated by the trainer. There was a new scar on her chest, in fact, that hadn't been there when she'd trailered off with him. She was bred already. We'd see what her foal looked like, and how it acted. For now, she was affectionate and beautiful.

Turns out, her foals were sensible and easy to train. She had a ten-year career as a productive brood mare ahead of her, and that is not a bad gig.

***

As spring approached, I got bigger and bigger. And so did Jane, who was pregnant when we bought her. She was bred to Ron and Deb's stallion, Woody, whom we'd admired when we were at their farm. Jane's belly expanded. I had to let out her quarter straps to get the harness to fit, and as I did so, my own belly bumped into hers. She turned her head back to look at me, her dark eyes seeming to understand our mutual situation. "So," I asked her, "got a name picked out?"

We hoped she'd have a filly. Most livestock breeders want female babies. You don't need as many stallions, bulls, and rams as you do mares, cows, and ewes. Male stock is usually gelded or sold for meat. We wanted a filly because we wanted to grow our herd. Jane's baby would be a start toward that either way, but if she had a filly, our herd could grow faster.

As for Richard and me and our baby, we didn't care. I had a feeling we were having a boy, but how could I know? I'd had an ultrasound early—to rule out twins because of how quickly my belly was growing—but I'd asked specifically not to know the baby's sex. Later, that evening, I popped the VHS tape of the ultrasound into the VCR to show it to Richard, who had not been at the procedure.

We sat dutifully on the couch, waiting. The darkness turned to static. The static sometimes arranged itself into baby-shaped blobs. It was kind of like a lava lamp. Toward the end, the baby's sweet, perfect hand emerged from the fog as if it were pressing against the screen.

"All right, then," we said. I ejected the tape and stuck a label onto it. Someday, maybe, we'd watch it with the very child it featured.

***

"Mama, when will ours baby come?" Marian asked me. The snow was gone. It was spring, but the point in spring when the earth looks as if it has undergone chemotherapy. Last year's vegetation is rubbed away, and there's no sign there will be anything to replace it.

"Our baby will come when the tulips bloom," I told her. I was kind of hoping it would be a while still—I'd been so anxious to have Marian, but now, awaiting this baby, I realized how much easier it is to take care of a baby while it is inside, rather than outside, your body. I wasn't quite ready.

Jane's belly was swelled taut. Her udder began to swell. We watched it closely. Ron and Deb and all the books we'd read told us that the mare's udder would begin to fill, and then at some point, the tips of her nipples would develop a waxiness. And once that happened (and it didn't always happen), we could expect the foal at any time. But you could also expect the foal at any time anyway. Kind of like a pregnant woman.

Day by day, though, Jane's udder grew. You wouldn't think a mare's udder is so big, but it took quite a while for it to fill to the point that the nipples stood out. And then one day, it looked like condensed milk had dripped out and clung. It looked like wax. We could expect a foal soon!

We took turns getting up to check on her all that night. In the first light of the morning, I heard Richard at our bedroom window. "Foal! We have a foal!"

I hauled my pregnant body up and into some clothes, then hurried out there with the foaling kit we'd prepared. Betadine for the foal's navel. Masking tape for Jane's tail. A thermometer to take Jane's temperature throughout the day. I had studied for this event!

In the paddock, Jane stood—alternately glowering at Jemima, who held herself respectfully at a distance, keeping a spruce tree between her and the older mare, and lovingly nuzzling her bony little baby, who wobbled adorably, searching and lipping at Jane's belly, wondering where this milk stuff was supposed to be.

How thrilling it was! Our first foal. And, Richard announced, "It's a filly."

She was perfect. Her still-wet coat was curly. Her face was narrow, her eyes were dark, her legs were those of a moose. They were so long; she had to work so hard to collect them beneath her in a coordinated fashion. Jane stood patiently as her baby searched up and down her belly, looking for the udder. Over and over the filly tried that area just behind Jane's front leg. I think that's where elephants have their udders, in fact, and it makes a lot of sense. I've seen dozens of foals look for their first drink, and that's always one of the first places they check.

But alas, the udder is well placed up under the mare's belly. And finally, almost in spite of herself, the filly found it and latched on.

"Kind of a sloppy eater," Richard noted dryly as she sucked and slurped.

"Marian!" She'd be awake any minute. I went to get her.

"Dat is ours baby?" We couldn't blame her for being confused. Once we'd cleared it up, she remained fascinated, watching the filly nurse, watching Jane lick and nuzzle at her baby's back and hip, laughing when the filly finally, after trying to do it carefully, just flopped down onto the ground for a nap.

We treated the navel and took Jane's temperature. We decided to name the foal Stella, which had been under discussion, it having been Richard's great-grandmother's name, as a possible name for OUR baby, should it be a girl.

Foals grow so fast. Newly born, they are a skeleton covered with hide. Their hocks swivel when they walk. By the end of the first hour or so, they are trotting—however bewilderedly—next to Mom in the pasture. Next day, they've begun to fill out. Stella followed the formula, amusing us enormously when she started running circles around Jane, bucking rebelliously, by the end of her third day.

"Her funny," Marian observed fondly.

"That's for sure," I agreed. Walking back to the house, we passed one of our flower beds and I saw, with a bit of a jolt, that the tulips were about to bloom. I hoped Marian had forgotten my promise about that. I still had a week to wait, and the baby was sure to be late.

"I'm going to plant onions tomorrow," I told Richard at dinner. "I bought some sets. I'll get the potatoes in next—they can take a freeze."

We were looking forward to the spring. Baby Stella complicated things somewhat, but we were used to accommodating little girls in our work. We'd already put a halter on her with a plan to teach her to lead. It's a common thing to see mares working with a foal tied to their side.

But morning found me in labor. Sporadic and unreliable, but I was leaking amniotic fluid and the midwife wanted me to come in.

We brought Marian to Cindy, her sitter, and drove in the glory that is early May in Minnesota to the hospital in St. Paul. It would be five years before I got onions in on time.

"Ours baby" was born in the evening, looking like a skinnier version of his older sister. It made sense—she'd been late, he was early. It twisted my heart, the look on his face. He looked as if he hoped he wouldn't be a bother, as if he knew he had not at first been very welcome.

Next day, Richard brought Marian when he came to get me. She gazed curiously at her new brother, then kissed his soft hair. "His name is Truman," we told her. "Kruman," she said.

Richard leaned down to kiss the baby as well, his hand gentle on the back of Marian's head. Apologetically he said, "Stella is sick." My heart sank. "She's limping pretty badly. I called the vet. He thinks it's something called joint ill. It's an infection foals get through their navel. He'll be out later to flush her."

Whatever that meant. All I could think of was how much we had already come to love that little sweet horse, how kind she was to Marian.

I was inside with the new baby and napping Marian when the vet came, so I didn't see how he treated the affected joint. Richard reported to me that the procedure went well and that we had to be consistent with the antibiotics now. Emily helped a lot with that. Only eleven years old and small for her age, she seemed to have made a mind meld with Stella. Emily would sit down near grazing Jane and nursing Stella, and before long Stella would come and sit across Emily's lap, where Richard would then be able to give the relaxed foal her shot of tetracycline.

(Just FYI—we stopped using Betadine on navels and substituted Nolvasan instead. I read that instruction in one of our breed communications. We never had another case of joint ill.)

Before long, Stella was up and hopping about again. It was beautiful to see her become rounded and full as the grass grew, the dandelions bloomed, and the apple trees blossomed. We felt like givers of life, like farmers already, as we surveyed our children and small herd.

But we were far from being farmers. All around us fields had already been planted with that year's crops. We were very late because of all this birthing.

"I'll drive them over to the field and start plowing," Richard told me. "Don't worry about this year. We're just learning. We're just trying things out. We don't need to make a crop."

I was always the worrier of the two of us. Richard stroked the underside of my chin one evening as I fretted about something. "This is how you calm a lizard," he explained. "Let's see if it works on you."

We had found a plow the horses could pull. A sympathetic bystander at the auction had explained to Richard how to adjust it. Leaving the kids with Cindy, we spent a morning trying it out with Jane and Jemima, making crooked-y furrows that sometimes disappeared when the plowshare jumped out of the ground. We learned as the horses learned. The furrows became more consistent. Putting two good furrows next to each other became less of an anomaly.

I learned that plowing with horses was a skill it would take some time to learn, especially in our rented rocky soil. We were using a sulky plow— that is, a plow one rides on. The seat is over the plowshare. A walking plow is a plowshare that has handles attached. The horses pull it through the soil as the teamster holds it at the correct angle. That is a terrific skill, one I have yet to master. The great ones walk along behind their teams and hold the plow handles as gently as I held my baby's hand. Soil parts before their feet. It is mastery.

***

"You're leaving now?" I was nursing Truman in the gray light of the early June morning.

"Yeah. I'll be home about two. Can you have the horses harnessed? We'll go plow for a while." Richard tucked his shirt into his pants, grabbed his briefcase, and kissed me and the baby before heading for the car. I hoped to get another hour of sleep, once Truman was finished. Marian would awaken with the sun, as usual, and our day would start with a trip to visit Stella.

It was a warm afternoon. The horses sweated, as we did. Stella solved the problem creatively—by then we could trust her to stick around without being tied to her mother's harness. She stayed well away from the plow. This particular afternoon, when she got tired, she simply lay down in the newly plowed furrow, where the earth was cool, and jumped up when I drove back and positioned the plow to make a new one. Then she lay down again, this time in the new one, nice and cool, watching interestedly as the summer day took place around her.

So now we had four horses. Only two of them could work. And the rules in our township were that no more than four horses could be kept on the size of lot we owned. So, no more horses. We planted potatoes in the part of the field that we got plowed, simply because they were the only crop that could tolerate being put in so late.

***

By Christmas of that year, we owned eight horses.

We'd bought a team of Norwegian Fjord mares, cute as buttons in their fluffy coats, and two yearlings, their bright eyes full of mischief. Just in case the Suffolks didn't work out—and because we lived in an area heavily populated with people who had Scandinavian heritage—we thought they might help us market our wares, once we had some.

The Fjord mares foaled. We had horses boarded in pastures around the area. It was very clear that we needed to move. We needed a farm. There was nothing around our current home that we could afford. We expanded our search across the river into Wisconsin. Sarah was graduating from high school that spring and planned to move west after that, and Emily had decided she'd like to try living with her mother, also out west. We'd miss them, but it also freed us up to look farther afield than the Stillwater school district for a farm.

"Let's give this a try," Richard said one Sunday afternoon, creasing the paper so I could read the small ad. "200 acres, River Falls." He set up an appointment to view it and another property nearby, and we set off within the week.

We'd seen so many places, there was no reason to think this would be The One—as I'd come to think of it. It was two hundred acres, no house, no barn, no buildings at all. I had no interest in building a house plus a barn plus all the fencing and so forth. But maybe the other place, which was already a farm.

We met the realtor and followed his truck up a steep driveway and then back to a rutted track. It was late winter. Marian was three years old, Truman nine months, and when we had gone as far as we dared drive the minivan. All I could see was white, rolling hills. Definitely not The One.

"I'll stay with the kids," I said. "You go." Richard hopped out and led the realtor up the nearest hill. I unstrapped fussy Truman from his car seat and clapped him onto a nipple. "Tell a story, Mom," Marian asked.

An hour later Richard bounded back to the vehicle, cheeks windburned, eyes shining. The realtor looked much less invigorated. "Want to see the next place?" he asked tiredly. Well, yes we did.

Richard revved up the van. Cold air sifted from him, but more than that, I could see that he was really excited. "So?" I said, strapping Truman back into the car seat.

"I love it," Richard said. "You can see St. Paul from up there. You feel like you're on top of the world. It's the best feeling I've had about a place that I can ever remember."

"But . . . no house?"

"We'll build one. We can build it wherever we want! We can build a barn and make everything the way we want it! We don't have to live with other people's dumb decisions!"

"We'll have to live with our own," I said. "And this might be one of them." I'd had my heart set on finding a place with a house and a solid, old barn. Flowers growing beside the house, big beams holding up the gambreled roof of the barn—I could see it in my mind.

By then we were following Tom, the realtor, down a hill to a highway. He drove across the highway and parked in the yard of a farmstead that pretty much fit my dream place. Except that the house was about twenty yards from the road. We'd spend every day with traffic noise and the worry of our children and animals getting hit by a car.

"We are not living here," I said to Richard. "That is today's decision." And he agreed. When Tom came to the window of the van, Richard told him we weren't interested.

"Yeah," he said. "After how much you liked the other place, I didn't think you would be."

It took a couple of months to convince me. I was so much against the idea of buying bare land that the day before we went to the bank to sign the papers, I took the kids to Cindy's and drove over to River Falls and puttered around the back roads, looking for likely places to buy. I found one on a hill, set back behind a row of tall pines. The house was small and neatly built.

There was a big, traditional barn. I drove up the driveway and parked. As I walked to the house, I passed a small outbuilding with tulips planted in a flowerbed beside it, and more tulips coming up in a flowerbed beside the house.

This was what I wanted. I had the feeling of home. Weird as it was, uncomfortable as I felt in doing it, I knocked at the door to see if the owner felt like selling.

The elderly woman who answered was half as tall as I am. I'll admit, my heart lifted a bit when I saw how old she was. Surely she was hoping to move to town?

"Sell? Oh, my, I never thought of it!" She laughed at the idea. The smell of banana bread wafted from behind her. "Why don't you come in and we'll talk? Would you like some coffee?" Anna Anderson. That was her name. She wouldn't sell me her farm, but she did ply me with banana bread, and told me stories about raising her family on the place. "We made a good living back then," she said, looking at me significantly. "You could back then, on a place like this."

I drove away, resigned to signing the papers I'd be presented with the next day. Richard and I would do as she had done with her husband. We'd make a place of our own.

The next day we signed for the loan on our farm. It was based on the value of our current house plus our income from our jobs (let's be honest—mostly Richard's). "I'd chase you out of here with a stick if you wanted this loan based on what you expect to make farming!" the banker joked, thus answering the question asked that evening more than a year ago with Ron and Deb.

"We own a farm!" Richard said, once we were in the car. "Us and the bank!"

It was exciting. It was scary. If you're familiar with the Muppets you will remember Beaker, who is Professor Bunsen Honeydew's hapless lab assistant. That was me—Beaker—aware in a foggy way that this might go very badly, but too overwhelmed to do anything but hold the bomb while the professor lit the fuse. But I smiled back at Richard. "Yup," I said. "Exciting!"

Chapter 4

We had to sell our house, which took five months and sixty showings, most of which I did myself. We had to build fences before we could move all our horses to the new farm. Turns out, we couldn't build the house wherever we wanted to—the township had ordinances about where we could build, and the county had ordinances about where the septic had to go. Even the bank had a say. The ordinances made sense, and I could see the bank's reasoning, but siting the house became a sort of Venn diagram exercise.

Even so, wheels began to turn. By October we had moved to the basement apartment in our nearest neighbor's house, on the same driveway. All our horses were in one spot, up on our hill farm.

We had some hay in a stack, covered by a tarp. The pastures we'd laid out were thick with grass. We expected the horses to eat from the pastures well into the winter. The snow cover would not be thick until at least January, we expected. Maybe it would be a light winter and they'd be able to paw through it even until spring. We'd supplement with hay. We'd haul water once a week and keep the tank thawed with our new electric tank heater. For now, things were good.

On Halloween morning the kids were so absorbed in their activity that I decided not to point out that it had begun to snow. Such an early snow—it would get them excited and then there'd be nothing to play in. Don't distract them, I told myself.

"Mama, it snowing!" Marian said before long. I looked out and sure enough, it was coming down in a businesslike fall. The ground was whitening fast.

That evening we took the kids trick-or-treating to our new neighbors up and down the road. Not getting stuck was the challenge of that candy run. In the morning, Richard switched the radio on from our bed on the floor. Our little apartment had one bedroom, which we had set up for the kids. We slept on the floor of the small "living room," which could also be called "the rest of the apartment." If the door had blown open in the night, it would have conked us in the heads.

Thirty inches of snow had fallen. The winds were fierce. It was the Halloween blizzard—still famous in the area. "Guess I'm not going to work today," Richard said. "I'd better go check on the horses."

If I could go back and remedy mistakes that we made, our lack of preparation for winter that year is one of the first I'd take up. Our poor horses were outside, in the wind, with little to eat. Morning and night for the next couple of days, one or the other of us would go up and use an axe to chop bales of hay out of that frozen stack—the tarp had blown off—and then drag them laboriously through the deep snow to the fence, where we flung them to the waiting, hungry horses.

The hay was of poor quality, I now know, but we put out a lot of it. Eventually it stopped storming, and we had a road plowed up to the stack. Richard had to leave on a business trip, and the plan was that I'd be able to drive the van up there with kids in it and feed the horses that way. Within a day of his leaving, the "road" had drifted closed. Instead, I hauled the kids on a sled up to the stack.

It was a twice-daily adventure for us, and I was lucky they saw it that way. Dressing them for cold weather became an art form for me. I dug the sewing machine out and made hats, mittens, and foot coverings that would not let the cold touch them. I needed them to be warm because I could never be sure how long we'd be outside. A fence might need fixing, or one horse might be missing and we'd have to hunt it down. I wanted them to feel part of our family's venture, so I had to make it all seem fun and exciting, and luckily for me, they made that easy to do. They were endlessly interested in being outside, seeing the horses, and helping me solve the problems that came up, however silly their ideas might be.

***

We didn't plan to live in the basement apartment forever. The loan for the farm included money to build a house.

One day the previous summer, Richard had taken the kids for a drive, giving me time to really think about what it was I wanted. Stillwater is a town full of beautiful, quaint houses. I'd been driving around it for years, thinking about what made me like one house more than another. Now I could put some of that thinking to use.

Well, except that the things that made those houses so attractive—porches and bay windows and so forth—were pretty much too expensive to add to my design. We had the money to build a basic house. I started by closing my eyes and imagining our life as we lived it indoors.

We'd be coming in from the car parked in the garage or, more often, we'd be coming in from working outside, and our shoes and boots would most likely be dirty. So, the first room would be a mud room. We'd need a place to hang our coats. Often we'd just be coming in to use the toilet—there had to be a bathroom right there.

One source of clutter was our mail. I wanted to keep that out of the rest of the house. So, an office would be on the way to the kitchen. I could then drop the mail in there as I went by with groceries.

The groceries would be plopped onto a counter in the kitchen. I'd make dinner and serve it from that counter, so the kitchen table should be adjacent to that counter. From there, we'd move on to living in the rest of the house. I didn't like hearing the television unless I was actively watching it. (We would not have had a TV at all if it were up to me, but Richard enjoyed it.) So, the TV would have its own room, with a door I could close.

In a short time, I had a rough drawing of how the first floor would look. Upstairs we needed three bedrooms, a room that could be a guest room or another office, and a bathroom. I wanted the laundry room downstairs, but it wouldn't fit. So that went upstairs as well.

Between Richard and me, we designed the house. And now it was being built. Every day the carpenters came and framed it up. As the winter passed, they closed it in. When the windows arrived and were put in their sashes, we could start to see how it would look. My sister and I as children had always admired houses with gambrel roofs. Richard's idea to make it a gambrel, with the roof on the north slanting down to include the garage, gave it the look of a barn. I liked that.

In the morning, I hauled the kids on the sled up to the haystack and dislodged a few bales to throw over the fence and spread out so the horses wouldn't fight over it. In the early evening, after Richard came home and could stay with the kids, I went up alone to do the same thing. I never went back down the road without looking at the dark shell of our house, imagining it finished and lighted, holding my family inside. It seemed impossible that this ever would happen.

***

"I have an idea," Richard said. "I was thinking—we should get a tractor. It takes so much time to plow a field. If we use the tractor for the heavy work, we can use the horses for the lighter stuff, like planting and cultivating."

"Aren't they expensive?" I asked. After all, that's one of the reasons we had all these horses in the pasture.

"This one I'm looking at is pretty cheap. It's old, and not too far from here. I can drive it home."

So Allis came to live with us—an Allis-Chalmers composed mostly of rusty metal and wheezy retorts. Richard continued to amaze me with his ability to figure out how to fix her and make her behave. Having a tractor, though, meant buying more equipment. She'd need her own plow, for example. And that was just the start.

In a very short time, we accumulated equipment necessary to put in crops of grain, potatoes, and corn. The amount of rusty metal on our previously bare land had come to some tonnage.

Katrina, who was rapidly becoming my favorite horse, was due to foal. She was a big mare—taller than Jane by about eight inches—and beautiful. I loved how she looked, and she was wonderful to work. I hoped she'd have a filly just like her.

We still didn't have a good place for a mare to foal. Katrina was on pasture. I walked up to the new house that night and threw a sleeping bag on the floor of what would soon be Richard's and my bedroom.

Outside, I saw Katrina's dark form grazing, and a sweep of my flashlight indicated nothing amiss.

After a short nap inside, I went back out into the April night. This time Katrina seemed restless. I sat down on an overturned bucket and waited in the dark.

She lay down, got up, lay down. Got up. What was she doing with her tail? I hated to use the flashlight, but when I did, I saw a tight, striped balloon coming out of her. It was nothing like what I'd been told to expect. Should I go get Richard? Was it worth getting the kids up, hauling them up the muddy driveway—too muddy for our van—in the dark?

I decided it wasn't. I waited. The foal was born, I saw with relief. Katrina turned to nose at it. Too much time passed. I finally ducked under the fence and approached.

In the light of my flash, I saw a perfect little foal lying there, panting. Making no effort to get up. Inexperienced as I was, I knew this wasn't right. I patted Katrina worriedly and ran down the muddy hill to tell Richard and call the vet.

We were lucky in our vet—Peter was young, curious, interested, and a good teacher. He got out of bed and drove his truck up to where I was waving my flashlight. He listened to my explanation and examined the foal. By then Katrina, this being her first foal, had lost interest and was grazing at some distance away. "Can we get this foal someplace warmer?" Peter asked.

The slippery little draft horse foal was harder to lift than you would expect, but we carried him to the house and lay him on the floor near the framed-in kitchen counter. We covered him with tarps to conserve his heat. "He's probably not going to make it," Peter warned me. "He didn't get enough oxygen during the birth process. From what you describe, the placenta detached too early."

There was no reason to keep Peter there to watch the foal die. I had him drop me at the house so I could tell Richard.

"I'll go sit with him now," Richard said. "You get some sleep."

In the morning, Richard came back down the hill. "I named him Lucky Romper," he said. "He's romping around somewhere else now. I waited till his heart stopped beating." Crying in his arms, I imagined my kind Richard's head against the foal's damp, cool chest, listening to the heart slow down and finally stop.

Marian and Truman were somber when I explained that Katrina's baby had died in the night. "But I wanted anuvver baby horse wike Stewwa," Marian wept.

So, the first important thing that happened in our house was a death. I wondered briefly if it augured ill for us, but I only had to think of the foal's brave efforts to breathe and the way Richard had stayed with him to the end—and I believe that foal honored our house with his death.

On the cement floor of the new house, Lucky Romper's moist shadow remained and was eventually covered with blue-and-white vinyl tiles. Sometimes I think of it there, next to our kitchen table, where so much of our family life has taken place. I hope we have honored him with our actions.

Chapter 5

We were doing the finishing work on the house as we could get at it. There was an unbelievable amount of trim wood to varnish. I'd made a friend, another Cindy, who sometimes took the kids so we could work, but of course we couldn't dump them on her every time we wanted to steal some time to spend on the house.

One time we bought a few hours of the kids napping on the floor in the mud room, window open to ensure fresh air, by my taking them to the Dairy Queen and letting them have French fries and a malt for lunch. The heavy meal—such a departure from our usual healthier fare—worked like a sleeping pill, and we got a lot done that day. In my defense, we only did that once.

A month before Truman's second birthday, we finally moved into the house. That had been a feat of will, involving deep mud and a neighbor's bulldozer hooked to a flat wooden sled, called a stone boat, on which we stacked our worldly goods. Eventually we had everything at least in the garage and were moving it slowly into the house and making the house into our home. At the same time, we were preparing to start farming in earnest. This meant plowing up a field so we could plant potatoes.

Richard and Allis accomplished that over a couple of days. Saturday morning found us among crates of cut potatoes—seed pieces—with Katrina and Lucinda harnessed and ready to pull the potato planter up and down the field.

It was our first planting on our farm. The old potato planter, which I soon dubbed The One-Man-Band, earned its name because of the tendency for the potato pieces to bridge in the hopper. What was supposed to happen was that they'd slide out in an orderly fashion, each one being caught and stabbed by a prong, then deposited in the furrow sliced out by a small foot in front of it, and then covered over by discs that were tipped toward each other and which left a small hill, indicating the row that had just been planted.

When it worked, it was great. But the maddening way the pieces hung up and were repeatedly stabbed, which mashed them, had to be addressed. I found that if I kept kicking the hopper as I drove the horses along a faint line left in the soil by a stick dropped down from the planter on its last time down the row, it often knocked the bridge loose, keeping the potato pieces dropping down as expected. It was a lot of multi-tasking.

"Good thing you're a mom and used to that kind of thing," Richard said admiringly. Yeah, I thought. Guess so.

***

After the potatoes were planted, it was time to start planning for hay. We were not going to have another winter like the past one, we resolved. We'd be ready.

Of the dangerous implements you can be riding on when the horses run away, the mower seems worst. The gnashing, clicking teeth of the sickle bar could cut through just about anything you might be wearing—especially skin. Even bone. It's so noisy, too, and the vibration is new to the horses. Ron and Deb had warned us to be wary of this.

"I talked to Ron about how to do it, and this is what he said," Richard said, settling beside me on the couch. "You hook the team to the mower, but then you tie them by the halters to a hayrack, and then you pull the hayrack ahead with the tractor, and then the person driving the team puts the mower into gear. If they want to run, they can't. They just have to get used to it."

By now, I was pretty much established as the teamster on the farm. Richard was so much better than I was at setting up and fixing equipment. He could look at how something was functioning and understand what needed to be adjusted. I was better at working with the horses than he was, and I think he felt comfortable, as I did, with the way this division of labor was shaking out.

"That sounds like a parade," I said. "But I like it. I'd hate to have a runaway on the mower."

We hitched Katrina and Jane to the mower, figuring Jane had experience with mowing and would be the calm one. It wasn't the ideal team, as Jane couldn't keep up with Katrina. But Katrina would learn from Jane, and then she'd be the calm one when we used her with Lucinda. And Jemima would have her chance to learn from Jane another day. We were always making those calculations.

Richard started the tractor and looked back across all the contraptions and horses to me. I nodded, my straw hat emphasizing my motion. He put the tractor into gear and moved forward slowly, and then faster—more like a walking pace. "Easy, ladies," I said soothingly as I put the mower into gear. The snicking of the sickle bar was almost as loud as the tractor, and the horses felt the added effort necessary to furnish the power to move the bar. Their ears went back exploringly. "Easy, ladies," I said. "Good job." Within ten yards they were moving calmly. Our parade carried on a bit farther than that, and then I motioned to Richard to stop.

"That seemed to go well," he said.

"Yeah, it did, I think," I said. "I guess we'll find out now."

We were at the head of the field we needed to mow. Richard untied them from the rack and moved the tractor and hayrack out of the way, and I clicked them forward and put the mower into gear. We were mowing hay!

The bar chattered beside me. The grass—at least a couple of feet tall—was a sheer wall, which, once the bar touched it, sank and tipped, falling over the bar to lie, slightly raised from the ground by the stubble, exposed to the wind and the air.

We'd learned at Ron and Deb's to make square turns. That technique came back to me with some practice. Sometimes the sickle bar clogged, jamming with fine grass and jerking hard on the horses' necks. Katrina, alone among all the horses I ever drove on the mower, learned to recognize the sound of an imminent clog and she responded by speeding up, which often avoided a clog. It was she who taught me to do that, in fact, as time went on.

If you have a good team, and a good mower with a well-maintained sickle bar, and if each time you stop to rest the horses you get down to oil the bar, and if the grass is not wet or very fine—we call it June grass, the fine stuff that is so good at clogging the mower—and if it isn't horribly hot and humid, and you know your kids are safe and happy . . . then I would say mowing is one of the great pleasures of farming with horses.

The smell is of cut grass and sweaty horse. When you stop to rest the team, the sound is of horses catching their breaths and birds singing. You step off the machine, always careful to stay behind the bar, and use the oil can to run a line of oil up and down the shiny metal teeth. You use the time to look at all the connections between the horse and the machine—are the heel chains fastened to the evener correctly? Is the evener free to move back and forth? Walk up, on the side away from the bar, still holding the lines, and talk to the team, so dark with sweat, their nostrils wide as they suck at the air.

When I was growing up there was an older couple in the neighborhood, Tom and Ruth. Tom milked a small barnful of Jersey cattle, and he farmed with a team of horses, hiring neighbors with tractors to do his heavy plowing for him. He was the last farmer in the area to use horses. I remember running past his place one midday when he was out mowing. We chatted as he rested the horses.

"Listen!" he told me. I did so, but I couldn't hear whatever it was he wanted me to hear.

"That's it," he said. "The quiet. I hear birds. I can hear you breathing hard from your run. If I had a tractor, we wouldn't even be talking! I wouldn't have stopped."

Farming with horses requires stopping to rest them. I often thought it would be useful to be a smoker. I'd look forward more to the rest stops. As it was, I was often pressed for time—the kids were with a sitter, and I had to get back to pick them up. Marian had a swimming lesson, and I had to finish this field before taking her. The horses could not work without stopping, and they needed to rest adequately. We learned to figure that into the job and our time calculations. We learned that it was better to use more horses on it than an implement required, often, so fewer rests were needed. God knows, we had enough horses.

Tom, my old neighbor, had learned to simply enjoy the pause his horses needed. Sometimes I was able to do that, too. The view from the fields was always calming. It was good to be away from the whining and bickering of the kids; I have to admit, I do remember that. If I was reading a book that was small enough to fit into a pocket, I took that along and read a page or two while the horses panted themselves back to being ready to do another round.

***

Mowing the High Field

The distant clouds have turned a dark blue-gray.
They seem to browse, while here I mow the hay.
My horses bow their necks to pull the clacking mower
And in the west the smoky clouds get lower.

There's sun on me, my team, the nodding grass.
Far off the storm clouds gather, loom, and mass
Like mastodons, with ancient, shaggy backs.
They stutter light. Turn gold, then back to black.

The stunned air shifts and shatters loud command,
Rain crooks a slant between dark clouds and land.
I rest the team, I oil the old machine,
Watch marsh hawks skim the rippling gold and green.

More distantly, wars darken other lands
Why for me this sun, grass-scented air, full hands?

***

Suffolk horses aren't that common. So, when we heard of one for sale, or going up for auction, we checked it out and often ended up buying it. One day, we read a classified ad in the agricultural newspaper we received. It was for a Suffolk mare, pregnant, with a foal at her side.

I called the number and spoke with the seller. His accent was hard to make out, but he said he had papers for the mare, and in both the case of the standing foal and the prospective foal, she'd been bred to a registered stallion. It was a long drive to southern Wisconsin. Before making arrangements with Earl—the guy we hired to haul horses for us—I called the ASHA secretary to double check. Her records confirmed what the seller had said.

So it was that Earl and I drove eight hours to Lake Geneva, Wisconsin, to check out a mare and foal. Earl was an older man with a jaunty red hat—I think it could be called a porkpie. He sucked on cinnamon-flavored toothpicks, a habit he'd taken up to help him stop smoking after a heart attack.

"This must be it," he said, pointing to a tilted mailbox on a weathered oak post. I'd been reading off the directions the seller had given me. Yes, the fire number matched. We eased around the corner and up the driveway, hoping there would be space to turn around up there.

"Gotta be," said Earl. "It's a farm, right?"

Well, I wasn't exactly sure, but I figured so. Now that we could see the buildings, yes, it was a farm. Neglected; had seen better days. But there was an old barn, several outbuildings, and a neat white house. Someone had taken pride in this place, and was still trying.

The little man met us, waving at Earl to drive the truck around the looped driveway so he'd be facing out once he was loaded. Earl gladly obliged. Truck drivers with a trailer behind are always happy when the exit path is an easy one.

I hopped down and met Martin VanHuisen, as he introduced himself to me. He was small, thin, and bent. There was no mistaking the sharpness of his eye and mind, though.

"Come see her!" he said, motioning for us to follow. In a small paddock, a chunky mare with a white blaze grazed contentedly. Her stocky foal had the same blaze and trotted up to the fence to be scratched by his friend, Mr. VanHuisen. And by me. Why not? he seemed to say. You've got hands. Use them!

We laughed at his boldness. "He is a character," tssked Mr. VanHuisen.

We went into the house to sign papers in exchange for the cash I was ready to hand over. The little farmhouse was as well-kept as the farm, but signs were clear of how difficult it was becoming for the occupant to maintain the standard he liked. Grime around the cupboard pulls, dust in the corners, a stack of mail, pill bottles in the center of the table, a list of phone numbers written in large, dark writing and placed prominently.

"Coffee?" Hard to believe it now, but I didn't drink coffee at that time. Earl did, though, and he had a cup from the reheated pot on the stove.

"You've got a nice place here," said Earl.

"Yes," said Mr. VanHuisen. "I came here from the Netherlands when I was a young man." So that was the accent. And the tidy farm.

"I had . . . ideas," he said. "I wanted big. Netherlands is small." A sip of coffee. "Sometimes I think, though, maybe I should have stayed."

He met my eyes, probably used to the attitude of people born in America and how we always think immigrants never regret leaving their homes. "But here I am," he said. "Fifty years."

Outside again, he showed us the sled he'd built, and the cart. "But now I am not strong enough, or steady, to use them," he said. "I am not long," he smiled wryly, "for this farm."

We loaded the mare, Gidget, with her jaunty foal, into Earl's trailer. Earl started the truck. I looked at Mr. VanHuisen. "What's the foal's name?" I asked.

"That is for you to decide," he said. "I think he will be a good horse for you." I shook his hand and he held onto mine for a moment, seeming to search my face before he let go.

"He'll have a good home," I promised. "Gidget and the foal and the new foal. Don't worry about them."

"I believe this is true," he said. "I am glad we met."

In the truck, back on the road, Earl finally spoke. "Good man," he said.

I tried to explain Martin VanHuisen to Richard when I got home—the sharp blue eyes, the kind face, the work-gnarled hands, the bent frame, the tidy but failing farm. He understood, it seemed. "Let's name the colt after him," he suggested.

I called Martin VanHuisen to ask his permission. He seemed bemused, but he approved. We registered Gidget's foal as Martin VanHuisen. Mare and foal joined the herd.

***

In some ways, Richard and I were well suited to each other. Both runners (he'd run about thirty marathons to my three, plus several ultramarathons), both early-to-bed/early-to-rise people, both sort-of vegetarians. We still ate poultry and fish on occasion. When we moved onto the farm, one of the first things we did was order chicks. I loved the idea of the kids collecting eggs, I'd heard that free-range chickens on a farm kept the tick population down, and it would be good for us to produce our own meat if we were going to eat it.

I picked up the peeping box at the post office. I wasn't the only one there to get my chicks. The workers behind the counter wore little, pleased smiles and it was easy to hear why—the boxes were stacked behind them and the little chirps lightened the industrial air of the room. Marian and Truman, important with responsibility, carried the box between them out to the car.

It was far too cold to put the wee fluffs out in the barn or even the coop Richard had built for them. We set up a heating lamp and makeshift cardboard-box pen in the downstairs bathroom. The kids were beside themselves for Daddy to come home from work and see the baby chicks. They nearly fell on the tiny flock in their excitement when he came in.

Truman spent most of his time with Marian and me. His true love was his dad, though, and at some point he figured out that he and Dad shared an important anatomical feature that Mom and Marian did not have. This was thrilling to him.

It was convenient to his toilet training that the chicks were in the bathroom. I could just remind him to use the toilet when I thought it was about time to do so. We were all admiring the chicks, Truman was on the toilet after peeing, and Marian was telling us something. Truman, suddenly inspired, interrupted her. "Hey! Hey! Wait! I haf to show the chicks. Hey, chicks, I haf a PENIS! I haf a PENIS!" He struggled down from the toilet to wag it at them.

Marian, our practical girl, looked at the chicks and then at her brother. She tilted her head and said, ego-piercingly, "Truman. They don't care."

The heat lamp went out overnight and three of the poor chicks were chilled enough to die. Glad the kids were still not able to count very high, I threw the little bodies outside into the tall grass before Marian and Truman came down to check on the babies.

We were able to keep the remainder alive, though, and they grew quickly. Soon they were out in their coop, and soon after that they were busying themselves around the yard. The Cornish Cross chickens were meat birds, bred to grow fast and produce a lot of breast meat. I've been told that they are so front-heavy that they can't breed normally—not sure on that one, but I don't doubt it. Within a couple of weeks they were half grown, it seemed, tall and gawky as misfit teenagers. When I went out to feed them, they raced at me, their heavy dinosaur legs lifting alternately and tipping their white, ungainly bodies this way, then that.

Before long it was time to butcher them. Richard pounded two nails partway into one of the logs of the corral. He fitted a chicken's neck between them, pulling back slightly on its legs so that the neck stretched out. He chopped down with the hatchet and that chicken's head went one way and its body the other. He dropped it into a barrel, where it flailed as it died.

I was ready with very hot water, into which I dipped the headless, dead chicken. I'd done this with my mother, back on our home farm. Not my favorite job, but I could do it. The hot water made it easier to rub away the feathers. It's called plucking the chicken, but first you rub. Then you check your work and pluck out the ones you missed. Then you hold the pimply body over a flame and burn off the stubborn ones.

Then it was my turn to behead. I picked up the chicken and felt its body against my midsection. I notched the chicken's neck between the nails. I looked at the chicken's clear yellow eyes, fiercely taking in what would be its last view of the world. Then I chopped its head off.

Once we'd killed the ones we wanted dead, we butchered. Neither of us was enjoying the day's work. Usually we talked and made plans and shared ideas as we worked. Doing this, though, we just went through the necessary motions. Horses, potatoes, wheat—that was life. This was death. The smell of it hung in the air.

Weeks later, when I took a carcass out of the freezer and thawed it to cook for dinner, the smell of it brought me back to that day. That wasn't the day we stopped eating poultry—we either ate or gave away the frozen chickens we had left. We even traded live laying hens for a few already-butchered chickens once ours were gone. One day as we were eating chicken, Truman made the connection between the living chickens outside and the dead one on his plate. He pushed it away. And that was the day we stopped eating poultry.

The kids and I loved the hens. They made their soothing clucks, they scratched in the dust or the grass for bugs, and best of all they laid eggs. They each picked a spot for it, and once we learned it, we usually found an egg there from that particular hen. There is something about reaching into a gap between hay bales and closing your hand around a warm egg that is so pleasing; it colors the rest of the day with a mild tint of wonder. Marian, Truman, and I spent about half an hour a day searching for eggs and would emerge from the barn with hay and cobwebs stuck to our hair and clothes, each of us with two or three eggs, all of which we could have purchased at the grocery store as washed and chilled little orbs for less than three dollars a dozen.

***

"I'll go to this auction on Saturday," Richard said, pointing at the newspaper. I read the line he was pointing at—two Suffolk yearling geldings would be coming up for sale. They were delivered on Monday of the following week, and I named them Gibbous and Crescent because of the moon-shaped markings on their foreheads.

The herd did not take kindly to them, I am sorry to say. The horses at the bottom of the pecking order are often picked on, and in this case poor Gibbous and Crescent soon learned to slip under the fence to get away from the bites and kicks of their herd mates.

I'd see them outside and dash out to grab them—it was such a regular thing to see them out that I left their halters on. One day I noticed they were out, and I had a few things to take care of before I could get to them. When I'd finished and went to put on my shoes, there they were on the step, peering through the window of the door into the mudroom, like a couple of Jehovah's Witnesses. I called for the kids to come and look, and we laughed at their inquisitive eyes, their unruly forelocks, their genial, curious expressions.

One day, though, they were out and I made an unexpected move as I reached for Crescent's halter. He jerked away. I felt a sharp pain, and when I looked down, the middle finger of my left hand was normal up to the middle knuckle. The upper two sections of my finger were at a right angle to it. I was home alone with the kids and had to get the horses caught, tighten the fence, make the calls for a friend to come and stay with the kids and to the clinic for an appointment to get my finger fixed, and then drive myself to town.

While idling in the room as I waited for the doctor to come and put things to rights, I wondered how it would be done. Grab and give it a twist, as in the olden days? Surely we'd progressed beyond that, I thought, imagining how that would hurt. Just then the door opened, two doctors walked in and did not miss a beat—they simply grabbed me, grabbed my hand, and *thwick*. The finger was straight again. So that's how it gets done. It hurt as much as I had feared.

***

Mowed hay had to be raked. We had a rusty old side-delivery rake with a seat on it; to be honest, it was terrifying to perch on that thing behind a team and put it into gear. The tines gnashed beneath and to the side of the driver. Jane and Jemima were our most docile team at that time, and I used them when I raked hay. If all went well, the hay poofed up in fat windrows, exposing the leaves and stems of the plants to the sun and wind. The sweet smell of drying hay wafted down to the house sometimes. "I wike dat smell,"

Marian said to me one day as we had lunch at the picnic table after I'd raked a small field.

"Me too!" declared Truman, sniffing emphatically. His eyes sparkled above his wrinkled nose. "You make your mama laugh," I said. He really did, several times a day.

***

"I'll be home around two today and we'll get that hay baled," Richard said as he stuffed his lunch bag into his briefcase the next morning before sunrise. "Shouldn't take long."

That afternoon he got the tractor started. We'd decided that even though it is possible to bale hay with horses if you have the right kind of equipment, we were a mixed-power farm. We used the tractor when it made sense. In this case, we had a heavy old Ford baler, and it made sense not to make the horses work that hard.

Richard had already been fiddling with the baler over the past few days. There were so many moving parts—everything needed greasing, loosening, tightening, figuring out. He started the Allis, to which the Ford baler was already connected. Like an astronaut initiating launch sequence, he made a series of moves, the final one being to slowly let up on the transmission pedal.

The baler stirred into action. The big arm reached roughly for a wad of hay. Richard moved the tractor, hauling the baler over the windrow. I jumped up onto the hayrack, ready to stack the bales.

Boom-da-CLANK! Boom-da-CLANK! The sound was familiar to me from my childhood and youth. Baling was pretty much all we did in the summertime on the farm when I was growing up. Dad had to make enough hay to feed the cattle all winter. The loft of our big barn had to be filled, plus the sheds. It was an enormous task.

I wasn't much surprised when the baler didn't tie knots for us. Balers are fickle machines. And you can't blame them. Tying a knot is a delicate process. Gathering up fifty pounds of hay and punching it into a dense bale is a job for a brute. A baler has to be a brute on tiptoe, I have always thought. I learned most of the swear words I know while watching my dad work on the baler. That this old bucket of rust was not tying knots did not seem unusual at all.

Yet again I was amazed at how Richard could look at something, fiddle with it, and figure it out. We couldn't talk to each other very easily over the noise of the tractor and baler—he'd geared everything down before he hopped off, but it was still noisy. I stayed close in case I could be of help, but

I doubted that would come to pass.

He took the nuts off the knotters and flipped them back and forth, cleaning out debris that had caught. He checked the twine that was feeding into the knotters—was it under tension, but not too much tension?

His ministrations finished, he hopped back on the tractor and tried again. The bales that emerged from the chute were floppy, but they were tied. We didn't dare make further adjustments.

It was a small field, but it took a few hours to bale it. The kids played at the edge of the field and in the car, which I'd driven up and parked in the shade with snacks and water bottles and toys in it. At the end of the afternoon, as the shadows were lengthening and the last of the unbaled hay was taking on enough humidity to make it tough, rather than dry and crisp, I pulled the last bale onto the wagon and added it to the fragile, shaggy stack. The counter on the baler read 210, meaning we'd baled 210 bales. I counted 140 bales on the wagon, which meant we'd baled 70 bales twice. They hadn't been tied, so we'd scooped up the loose hay as it spilled out of the chute and run it back through the baler. Just another modification. Just another day on our nascent farm.

I helped Richard fit a tarp over the fruits of our labors. He tied it down as I drove the kids back to the house and started dinner. They were so tired that Truman fell asleep in the bathtub—Richard sat on the edge of the tub, a towel spread out on his lap, and I put our wet weasel of a boy into it. "I tired, too," Marian confided to me as I dried her feet. "You worked pretty hard up there," I said, and she nodded gravely. No one needed a bedtime story that night.

Our plan was to grow and sell grains. We were in love with the look of a wheat field waving in the summer sunshine. We loved the idea of growing organic wheat—winter and spring—and spelt. And rye. Any kind of grain.

We had planted spring wheat with an old wooden grain drill. We got it in later than we should have. We knew that when we planted it, but we went ahead anyway. Something is better than nothing, we told each other. The potatoes we'd planted were up. We didn't have a good way to hill them other than with an old cultivator Richard fitted with wide sweeps. Sometimes I went up with a hoe and tried to pull dirt up around the plants. It was a lot of rows of potatoes—I only got about half of them done, try as I might.

By summer's end we had made about half the hay we needed for our herd. We made arrangements to buy the rest, having it delivered at regular intervals so we didn't have to worry about it getting wet. Our spring wheat was up, but we hadn't been able to harvest it. We'd hoped to find a small combine by the time we needed it, but nothing was showing up at auctions. The field was mostly weeds. Richard harnessed a team one Saturday morning and mowed it all down, then mowed it the other way to chop up the plants smaller, so they'd incorporate into the soil more easily. It would be easier to plow it without the long stems of the weeds and wheat plants. He ran a disc over it, too. The wheat seeds sprouted after the next rain and made a nice cover as autumn approached.

The potato plants were dying back. The kids and I often went up to the field with a wagon, buckets, and shovels. "Shubbles," Truman called them. "I shubble dis!" he'd exclaim, taking his little shovel and attacking the ground.

I showed them how to look for the dried stems of the potato plants. For a while they waited eagerly as I dug into the soil, unearthing potatoes. Then they plunged their hands into the hole, eager to find the buried treasure, in this case, potatoes. I worried I'd accidentally cut them with the shovel if I kept digging the hole wider and bigger. I took to digging up one plant, then moving on to the next as they scrabbled for the potatoes in the first hole. After they'd grabbed up the potatoes they could see in the first hole, they scrambled to the second—and I went back to the first to dig further and unearth more.

We did this on into the fall. One day there was a sudden explosion of flapping beside and then above us. We startled, and I had my mouth open to explain to the kids that it had been a pheasant, when I heard Marian's clear voice in the autumn air. "Dat's a pheasant," she told Truman. "Dat's a kind of bird." She and Truman stood like a couple of art lovers in front of a good painting, watching the near distance for another sight of the pheasant.

I looked down at my hands, so dirty from pulling potatoes out of the earth. The buckets of potatoes tilted this way and that, waiting for me to lift them into the wagon the kids liked to help me pull down the hill. Marian and Truman stood still, watching intently as they saw the pheasant one more time as it flew away. This is why we do this, I said to myself.

Chapter 6

Sometimes that's how I felt. But not always. Richard and I shared this farming dream. But five days a week he left for work, a completely different world, and I was left with the kids, horses, weeds, dirt, and what felt like mindless, thankless, work.

Sometimes when I was cold or hot or especially dirty or tired and still had to finish the job at hand, I remembered my other life. Because I had not enjoyed teaching, I'd gone back to magazine editing. Once I'd had an office, clean clothes, and smart and funny co-workers, and we put out a product every month that pleased us—and, more importantly, our readers. I'd left that to marry Richard and had moved back to Stillwater to be near him and his work, which was much more stable and remunerative than mine. I missed working with colleagues, corresponding with authors, editing manuscripts, evaluating artwork, having a life of the mind.

Now I worked mainly outside. My most frequent contacts were with my children and horses. Richard, on the other hand, was gaining prominence in an area called Quantitative Structure Activity Relationships. It was a way to predict the toxicity of molecules even before a substance was synthesized. It could reduce the need for animal testing, which I thought was most important. And it could save companies a lot of money in reduced need for testing.

He was traveling more than ever. To Washington, D.C., to Denver, Houston, Dallas, Raleigh. He went to Europe a couple of times a year, for at least two weeks each time. When he was in the United States, he called every evening without fail. It was more difficult when he went to Europe. I learned to be happy with a call every few days, quite often routed through his work friend Eric's phone.

"Hi, Maureen!" Eric would say after I'd picked up the phone and said hello. "I've got Rich here!" And then we'd have a stilted conversation, me quickly trying to convey some sense of whatever catastrophe was happening at the moment, and Richard telling me about the weird food the fancy chefs were concocting for their group.

That didn't go well, I can tell you that. Hard as Richard tried to be sure the fences were tight and the fence charger was working, or that the equipment I'd need while he was gone was set up correctly and well-greased—something was certain to go wrong. The horses would get out, the pitman on the mower would break, Truman would need stitches, I'd slam my finger in a car door.

I missed him to the point of tears, and then once he called, I hated him for not being home. When he did come home, he could do nothing right, either. That first spring on the farm, he had arthroscopic surgery on his knees. The dozens of marathons and ultramarathons and trail runs over the years had settled into bone spurs under his patellas.

"I'll be up and on an exercise bike within an hour of the surgery," he told me as we planned the week ahead. I was glad. It was a busy time, we were excited to get things started in the fields, and having him pain free was going to be wonderful.

I dropped him at the hospital for the surgery and took the kids to Tiny Tots, a class offered by the local Parks and Recreation department. I swung by the grocery store and then back to the hospital. After a short wait, the doctor came out and waved me into the recovery room.

"He's doing great," he said. "He'll be going home as soon as he's ridden the bike."

Richard didn't look great. I'd imagined, because of the breezy way he'd presented this procedure to me, that he'd just pop back up onto his legs.

"Hi," he said, weakly. "I'll be ready to go soon."

"Are you sure?" I couldn't quite believe it.

"Mrs. Purdy?" a nurse caught my attention. "I just want to talk to you about removing the drains."

"Wha—?" The nurse showed me the plastic tubing coming out of Richard's knees.

"Tomorrow, just remove them," she instructed.

"How do you do that?" I asked, genuinely mystified. Weren't they stuck on something in there? Would I wreck the effect of the surgery by pulling on it?

"Just grab and pull," she said. "They'll come right out."

The next day that is what I did. One came out easily. The other was more resistant, and a rope of blood clot slid out of it and hung from the incision against Richard's leg. "What do I do with that?" I asked him. He shook his head, also befuddled. I took a piece of gauze and pulled more clot out and broke it off, sort of. "You realize I don't know what I'm doing, right?" I asked him. "This is where modern medicine has taken us."

We'd expected him to be up and back to normal within a few days of the surgery. "Are you sure the doctor didn't mean it was an in-and-out procedure for HIM, not you?" I asked in frustration as Richard limped in obvious pain around the house.

    He felt bad about it, too, but there wasn't much to be done. Adding fuel to the fires of my frustrations, Truman began a period during which he could not bear to be anywhere but next to me. I had to hold him in my arms or on my hip, carry him on my back, or park him in his high chair within six inches of my body. If we were separated, his wails were epic. I carried ear plugs for those times I had to sever the connection between the two of us, for however brief a time.

    Marian watched interestedly. "It's wike you're the 'frigerator and he's the magnet," she observed. I had to laugh. That's exactly how it was.

    One day I had to go to town for some reason—errands, shopping, whatever—and I took Marian with me. Richard was home, still recovering, and Truman could be with him. It would be good for both of them, and a nice time for Marian and me to have together.

    We drove away, not looking back.

    Behind us, Truman caught sight of the minivan leaving. "Aaaaagh! Mamaaaa!" He ran to the door, flung it open, and raced after us in his stocking feet. He was fifty yards ahead of Richard by the time Daddy saw what was happening.

    "Truman!" Richard called. "Mom will be back soon!" He told me later that Truman never looked back. Just ran. Richard couldn't run. He had to walk on the muddy gravel driveway with its edging of slush and ice, following Truman, who finally exhausted himself and stood, weeping, bereft, left behind.

    Finally Richard caught up to him. "Hey, bud, let's go back. It's cold out here." Because of his fragile, recovering knees, Richard couldn't pick Truman up and carry him back. The two of them walked slowly back to the house, one of them weeping, heartbroken, and frequently putting up his arms to ask to be carried, the other frustrated and cold and sorry he couldn't oblige his son by carrying him. Neither was very pleased with the other, and somehow, when I got home, I was the villain. Funny how that goes.

***

When my sister and I were growing up on the farm with our five brothers, our mother worked as a nurse at the local clinic. We were left with the housework and cooking. Andrea didn't mind cleaning, and I didn't mind cooking. We agreed that as adults, we'd live next door to each other and I'd cook for her family and she'd clean for me (I somehow never saw myself as having a family). Now, as an adult, I still hated cleaning. I liked living in a reasonably clean house, though. I fitted cleaning in around the work of the farm. And I fumed bitterly if I felt that Richard was taking my work for granted or being cavalier about helping to keep the house clean.

I blew up at him, over and over. I wrote notes. "And when you DO help, why do you do such a half-assed job? Why can't you AT LEAST play with the kids instead of doing logic puzzles or watching TV?" I burned over these things. I'd come home from some outing, maybe even only to the grocery store, and find the house in shambles, toys all over the floor, dirty dishes on the table, pee-soaked underwear (Truman was not yet trained) on the steps. "What were you doing?" I'd ask, incredulous.

And he'd look back at me, equally astonished. "I was taking care of the kids!" To him, that was one job. Not one of several jobs that had to be accomplished coincidentally.

Another time I asked him why he hadn't vacuumed up a mess. "I tried," he said. "But the vacuum made this crazy sound, like an alarm, almost. I didn't know what to do, so I turned it off." This from a man who could wire and plumb a building, take apart the baler and put it back together, figure out how to repair the mower. But not, apparently, stick his finger into the vacuum-cleaner hose to dislodge the obstruction causing the problem.

One winter weekend before he would be leaving for a two-week work trip to Belgium and the Netherlands, he got up on Saturday morning and immersed himself in a book of logic puzzles.

I waved my hand between his nose and the page. "You're going to be away for two weeks. You can do puzzles on the plane. You won't see your kids for FOURTEEN DAYS. Is this really how you want to spend the weekend?"

To his credit, he shut the book and put it in his briefcase. He spent the weekend with us, engaged.

It was as he always said in response to my outbursts of anger. "Why don't you just tell me what you think before you let it all boil over?"

Because he SHOULD see the mess at the same time I see it, I used to think. Because he SHOULD care about the things I care about.

But of course he didn't see messes. Even if he saw them, they didn't bother him as they did me. This infuriated me. Seeing him respond to my comment BEFORE I got angry, though, I believe was the first step for me toward the better solution, which is communication. Many married couples figure that out pretty quickly. I did not, and if functioning at a slow burn can cause permanent inner damage, I probably have scorch marks on my heart.

***

Truman soon grew out of his need to be within inches of me at all times. He was such a skinny, wiggly boy that Marian and I liked to tease him that he was not really a boy at all, but a monkey who had jumped into the stroller when we were visiting the zoo. Real Truman was in the monkey cage, being raised by a monkey mother.

This delighted him. One day he did something silly and I said fondly, "You are such a silly monkey." He looked up at me, his face merry with the idea of being a monkey. "Watch out!" he warned me. "I might eat all your blanas!"

***

After the death of Lucky Romper, Katrina's foal, we worried over the impending birth of Lucinda's baby. This time, we'd both be present. We called Peter, our vet, to warn him that he might receive a phone call in the wee hours of the night again. We brought the kids to Cindy's house to spend the night.

"Now look, she probably won't foal," I said to Richard. "Murphy's law."

But Lucinda did foal that night. We had her alone in a paddock near the house. She didn't even wait till the wee hours, considerate mare that she was. She was huge—the largest Suffolk horse we'd seen at that time—and being pregnant made her even larger. It must have been about eleven that night when her water broke. She delivered a large filly within the half hour. We named her Margaret, and she was quick to stand and suck. Richard measured her against himself, remembering where his hand had bumped his ribs when he'd held it against her withers. She was tall, this little clumsy girl. Given the conformation and bloodlines of her mother, we expected great things.

Twenty-five years later, Richard, Truman, and I would kneel beside Margaret as she lay dying, not even two hundred yards from where she had been born. Marian was across the country, waiting to hear via the phone what was happening to our beloved mare. Neither of our kids could remember the farm before Margaret. Dick—not Peter, who had moved on to another career—administered the pink drug. The three of us held Margaret as she died, much as Richard and I had held her when she was born. She'd never left the farm in all her life. Truman asked to bury her. He spent the morning digging a hole with the backhoe. When I went down later, I found the area smoothed over, and a stone placed at the foot of the grave. I planted violets there, and Richard scattered clover seeds. Marian cried over the phone when we told her.

But on that chilly first night of Margaret's life, we could not, of course, know the future. We were just happy to have the first live foal born on Baldur Farm, and we were happier still that she seemed so fine. She would go on to fulfill every dream we had for her—but as I have said, we didn't know that then. All we could do was name her Margaret, after my dear friend Peggy, and make sure she had everything she needed.

Dinah foaled next—dramatically, as one would expect from her. She was a month overdue, for one thing, if Jim's dates had been correct. She began bagging up and finally, after about ten days, her udder was rock hard and little droplets of wax clung to the tips of her nipples. Surely she would foal that night.

We had her in a small paddock where we could watch her. I dressed as warmly as I could and went out to the sleeping bag I'd spread over a bed of old bales. It was a cold spring night. At one in the morning the wind came up. At three the sleet began. Dinah showed no intention of having her foal. At four I was beginning to freeze as the sleeping bag became wet. I went in, beaten by the weather.

At first light I went out and found Dinah with a small filly struggling to stand up. Try as she might, it was not going to happen. I didn't find the placenta, either, which was another cause for worry.

I called Peter, who came over and worked with me for an hour, trying first to get her to stand. We got her up and then tried to get her to suck. She was probably too cold to give it her best effort, sweet thing. She'd try all along Dinah's underbelly, stopping short of where she needed to latch onto, and then she'd stand back and shake her little head, as if to say, "This is really hard!"

Finally, she lay down, sinking my heart as she flopped onto her side. I felt as if we were going to have a repeat of Lucky Romper.

Peter milked Dinah—an act of bravery on his part given her proclivity to kick—and put a tube down the filly's nose and fed her the colostrum. She rested for a while and then got up, warming our hearts on that cold morning, and finally we were able to tip the nipple just right as she nosed in the correct area, and BINGO—she latched on. She sucked noisily, dripping milk from her whiskers, and Dinah reached back and nuzzled her backside.

"That stimulates the foal to suck more," Peter pointed out. Before long the filly lay down again, Peter tube-fed her the rest of what he had milked from Dinah, and then he reached into Dinah and found her placenta and pulled that out. What a hero he was that morning!

We named the foal Rosalie, after another of my friends.

Because they were nursing their fillies, we felt that Lucinda and Dinah needed grain every day. I went out before Richard left for work to put the two low rubber basins out and to pour the oats into them. They were always eager for their feed, and this particular morning were being fierce with each other, as if afraid the other would eat both their portions; neither of them was wrong about that.

I poured out the grain and turned to leave the pasture. At the same time, Cinda whipped around to kick Dinah, full force, with both hind hooves. Except that I was in the way and took the blow to my back. I was knocked flat. Was my back broken? My feet still worked, though I couldn't stand on them. The rest of the herd was coming, and I had no interest in being carefully examined by several draft horses, first with their noses and then, however carefully, with their pawing hooves.

I slowly, painfully, crept along the ground until I had gone under the fence. That was as far as I could go. Richard would have left for work by now. I'd have to crawl back to the house and be sure the kids were ok, and I could call for help if Marian would bring me the phone.

The pain sometimes subsided and then would spike, like a jab. But I had to get back to the house—just . . . it really hurt to move.

I heard Richard approaching. He had not left for work, luckily. "What's wrong?" He lay down beside me, which seemed odd—still does—and spooned up to my back, as if he could make me better that way.

"Cinda double-barreled Dinah, but I was in the way. Got my back," I said.

"I'll bet you have some broken ribs," Richard said. "Can you get back to the house?"

With his help, I could. We called our neighbor to stay with the kids and I went to the hospital, where we learned through x-ray that nothing was broken. The ribs had just been lifted from their sockets and then snapped back. "Hurts like a banshee," the doctor said, sympathetically. "And it will for a while."

With that encouragement, we went home, where I sat quietly in a chair, trying not to move any part of me. Sometimes some muscle would have an involuntary twitch and the pain would spike up, as anyone who has had damaged ribs knows, and I couldn't help but make a small yelp in response.

The kids were being good, playing on the floor around me (Richard had gone to work). After one of my yips, Marian stopped and sat back. "You do dat because of it hurts you, right, Mama?"

I had to recover fast; it was spring. I dreaded more than anything a cough or a sneeze. I learned to sidle rather than walk. It was often excruciating, but less so as the days passed. The bruises on my back were impressive, but they didn't tell such a dramatic story as the backs of my legs, which were purple. Lucinda's kick had knocked me down just as one knocks down one of those inflatable punching dummies. My knees hadn't buckled—I'd just flown forward, stretching the backs of my legs enough that they bruised from it. Even Richard, who is seldom impressed with anything, still remarks on that phenomenon.

Chapter 7

Our farm needed a barn. Margaret and Rosalie could have been born indoors, where we could have kept them warm and dry, not to mention it being easier on the person watching for the foal to be born. We needed a place to store hay and to keep our equipment out of the weather.

The Fjord horses were more valuable than the Suffolks. We preferred working the latter, though, because they were larger and could do more work. We loved the Fjords for their personalities and willingness to try anything—but we needed that barn.

We decided to sell Bibi and Kate, the two Fjord mares. Bibi's filly, Inga, was a yearling, and we'd keep her. Kate was about six months older—I loved Kate especially, but . . . the barn. This was my first experience with selling horses I loved, and I have to say that I've never gotten used to it.

They were easily sold, and with that money we had the barn built. It was a three-sided pole shed and went up in about three days that October. We also had automatic waterers installed. This meant having an excavator dig trenches six feet deep from the well up to each end of the barn. He'd no sooner finished than we had three days of steady rain. The trenches filled with two feet of water. Nothing could be done till they drained—which, given the heavy clay soil, would not happen soon. I kept picturing the kids falling in and drowning; for two weeks they were not allowed out of the house unless I was with them and able to pay full attention to where they were and what they were doing. I think I'm probably giving something away here with regard to my usual level of parenting—normally I paid casual interest to their activities as I pursued my own. But until the pipes were laid and the trenches filled, I was a model mother. My kids had two weeks of that and were itching for it to end.

That winter we did not have a fierce blizzard. We had snow and cold, but not in remarkable amounts. It would have been nice to have had such a winter the previous year, but that was not how it worked out. The season was bearable. It was so much easier, too, to live so near the horses. I didn't have to bundle the kids up to pull them on the sled up the hill to the farm each time I fed the horses. Most mornings I positioned them in front of the TV, to be honest (I wish I were a better person and could tell you otherwise) and ran out to feed the herd. I could do it in the first half hour of Sesame Street if all went well, which it usually did.

"Maybe they have tumbling classes for kids," Richard suggested one evening, watching Marian hop from one vinyl square to another on our blue-and-white floor.

I checked, and indeed, Parks and Rec offered them at the local university. Marian enjoyed her "stumbling" classes, as she called them, but never as much as her parents enjoyed that she called them that. We loved her so much, but she'd inherited from each of us a gene for clumsiness that made "stumbling" the perfect adjective for her performance in class.

Truman, on the other hand, could climb anything and then balance on it for as long as he liked. Richard took Truman with him to a farm auction one afternoon. Truman amused himself by climbing on a piece of farm equipment. "You shouldn't let him do that," the man next to Richard admonished him. "He'll fall."

Richard weighed that possibility against the effort that would be involved in keeping Truman off the equipment. "No," he told the guy. "He won't." And Truman didn't. He never did.

One January morning Richard fed the horses before he left for work. I didn't have to get dressed right away, which seemed to me a great thing. I slipped on a long, heavy flannel nightgown and started putzing at little jobs around the house that I'd been neglecting. Marian was impressed with my attire and decided that I was a queen. She picked up the back hem of the ratty nightgown and followed me reverently, holding it up in her two hands as I put away laundry, took Truman to the bathroom and wiped his wee bottom, swept the bathroom and cleaned the toilet, all the while getting colder and colder on my backside.

All her questions were directed to me as Queen, as if that were my name. "Queen," she asked. "When will you make breakfast for us?" I thought that didn't exactly jibe with the role she'd assigned me for the day, but of course I fed them.

***

"You look like you've been four-wheelin'!" the carry-out boy said, sliding the grocery bags into the van, snugging them below the kids' booted feet. It was true. Our white minivan was plastered with mud from the driveway. Springtime equals mud time on a farm. Richard and I were anxious to get into the fields. We hoped to get a crop of wheat this year.

"If the price of organic spring wheat is xx/bushel, and we get xx bushels off that field, we can make this much," Richard said. "But if we mill it ourselves and sell it as flour, we'll make that much more."

Value-added was the buzzword of the times for small-scale farmers. We all knew we couldn't compete with the big farmers if we sold our crops as commodities on the same market they did. Their efficiencies of scale were so much better than ours.

"No one takes into account the cost of what gets done to the soil," our new friend Prescott said one day. We'd met him and his wife, Juliet, when we'd heard about the farm store he was opening on their place some miles from ours. We drove out to look it over and I bought a shuffle hoe that is still my favorite garden implement. We soon found ourselves deep in conversation and making promises to get together for dinner soon.

"It's cheaper to buy inputs from somewhere else and truck them to the farm and use chemicals to control the weeds and bugs. It's like they use the soil just to hold up the plants," Prescott complained. "If you want to work with nature and actually build the soil and keep the water clean, you take a hit."

"We're working on our organic certification," I said. "We'll get a better price with that label on our crop."

"That's true," agreed Juliet. "But it won't be enough to offset the cost of growing it, at least at first."

"She's right," said Richard. "Until we figure out how we can be really efficient, we'll waste money and energy getting a crop. But that's going to be the cost of doing business." We were lucky—he was keeping his corporate job and we weren't depending on the farm for an income, at least for a while.

"I think your best bet is milling the wheat yourselves," Prescott said. "I think you're on the right track there. And then make it into bread and cookies—vertical integration!"

More buzzwords. We talked and talked about our dreams and how we hoped to change the way food was grown in America.

We were a long way from needing a mill, though. First we'd see if we could grow wheat.

It had to be planted by April 15, we learned from our extension agent. Every day after that would result in less yield.

"It's just too muddy to plow," Richard said, returning from the field. It was April 10. The snow was melted as of a week ago. "That clay will take forever to dry out."

He was right. A ridge of clay runs across our township, and our farm is in its path. Down the hill from us, neighbors were merrily plowing their lighter soils. We were bogged down.

By April 20, Richard had the field plowed and disced. We hitched Jane and Jemima to our grain drill and began drilling in the wheat. It was a six-foot drill, meaning that we could plant a six-foot-wide swath. I drove the team around and around the field, stopping to rest them while Richard dumped in more bags of seed wheat.

I used Katrina and Lucinda to pull the cultipacker around the field, further disrupting clods of dirt and pressing the soil around the seeds. After a day of planting and a day of packing, we had a wheat field.

"I can already see the weeds coming up," I said, pointing out a few cotyledons that were too broad to be the sprouts of wheat coming up.

"Velvet leaf," Richard admitted. "That's a problem we're going to have. Getting into a rotation will help."

Around us, the conventional farmers planted corn, then soybeans, then corn. My own dad had planted corn, soybeans, and alfalfa in a rotation. Rotating crops was an old idea. It disrupted the cycle of pests, each crop used different nutrients, and the beans and alfalfa added nitrogen to the soil, which kept the cost of fertilizer lower. We'd decided on a four-cycle rotation: Potatoes, wheat, alfalfa/vetch, and buckwheat. Richard was already plowing up the other fields. The Allis was doing a good job for us, once he'd figured out the correct amount of tinkering that had to be done each time he wanted to start her.

Planting potatoes meant getting out the one-man-band again, that big rusty planter that required so much of my coordination. We'd ordered several hundred pounds of potatoes, and I cut them into seed pieces over a few afternoons as the kids played in our new barn.

I used Katrina and Jemima this time. I used to think we had to use horses in set teams. Because our pasture was so big, though, I soon developed the habit of taking whichever broke horses were closest to the gate. If they had worked hard recently, of course, I gave them a break, or if I needed a particularly quiet team, or if I wanted to work a newly trained horse, well, then I'd seek out what I wanted. But mainly in those days I was tired, rushed, and eager to get on with the task at hand. So because Katrina and Jemima were standing where I could see them, they got the job.

The sun was barely up that Saturday morning in May, my favorite month of the year. Richard stayed in the house to nudge the kids awake as I went out to catch and harness the horses. He came out with the kids, each one holding the little bag of breakfast food I'd packed for them the night before, and had them "help" him pull the potato planter into position. They strained mightily to swing the pole outward, making it easier for me to step Katrina over it once I'd gotten the team harnessed and ready to go. Richard lifted the neck yoke up and snapped it to the breast straps on the horses. He fastened the tug chains as I held the lines and made sure the kids were well to the side of the whole operation.

It was nice having Richard around when I hitched to an implement. It was safer, for one thing, and I liked having his opinion as to whether the chains should be a link tighter or not. Working together was always better than singly, but because of his day job, I didn't have that luxury very often.

I walked behind the implement as the horses hauled it up the hill. Richard drove the pickup, and the kids were allowed to ride in the bed of the truck, seated on the floor and surrounded by the crates of potato pieces.

Our farm on a beautiful May morning—there is no better place, in my opinion. Up on the ridge, where the potatoes were to be planted, we could see the high buildings of St. Paul, thirty miles away. Birds sang, the grass sparkled with dew, the field was dark and fudgy looking, and the horses stood nicely as Richard filled the hopper of the planter.

I swung onto the seat and began our day's work. It was cool, but as the sun rose I could feel heat—both from above and also radiating up from the dark earth below. The horses began to sweat; I could smell it before I saw the darkening on their sides and legs.

The kids played in the cab of the pickup, at the edge of the field, out on the farm road, and on the overturned crates as Richard emptied them into the hopper. You may be expecting catastrophe to strike at any moment, but it did not. We had only the frustration of the potato planter and its idiosyncrasies to make the day less perfect. Sometimes I went a couple of yards without a potato dropping into the furrow—despite all my kicking of the hopper, agitating with one hand as the other held both lines, nothing seemed to prevent the seed pieces from bridging—that is, several potato pieces in the hopper pressing against each other and creating a kind of arched bridge that prevented other potato pieces from dropping down to be planted.. Even worse, sometimes they'd bunch up at the exit from the hopper, and the prong that was supposed to stab each piece and place it in the furrow instead mashed a bunch of the pieces.

It was wasteful in a few different ways, and by the end of that day we were determined to find a better way to plant potatoes. Yet we were grateful to Jemima and Katrina for having been so patient as we'd stopped, started, and stopped again, trying to make the planter work better. The mares were barely four years old. It wasn't as if they worked every day. They were just good horses, and when I put them back in the pasture that day after taking their harness off and rubbing them with handfuls of hay, Marian and I watched them drop to the ground and roll, roll, roll.

"That feels so good to them," she observed. I agreed.

The bathwater turned brown quickly that evening when Marian and Truman had their bedtime cleansing. I let it out and ran some more. Once they were cleaned up, they fell asleep with barely a few words of a story. Richard and I walked back up to the field to see it before the light was completely gone. Our potato field looked like a highly magnified swatch of corduroy, the hills of potatoes regularly spaced, following the contours of the field's uneven surface.

"That looks so nice," I sighed. I felt as if this day was the first to approach what I'd hoped for when we decided to farm, and to farm with horses. It had been a long time coming.

***

Auctions took a lot of Richard's time away from the farm. We were lucky to live in a part of the country that still had horse equipment in fairly good condition. The fields in our area were small, compared to, say, fields in the Red River Valley where my parents had grown up. My mother's dad had farmed mostly with tractors, using horses only for milk delivery. He farmed whole sections of land. Farms in that kind of country had moved quickly to tractors and larger equipment.

In our part of Wisconsin, on the very edge of what is called the Driftless Region because it has never been glaciated, the hills and woods kept the fields smaller. It still isn't terribly unusual to see a farmer on a tractor that could, if it were a person, collect social security. Horse-powered farming lasted longer in this area, so Richard watched the auction bills for what would be on offer.

Most auctions today take place online. Back then, though, they were an in-person kind of event. I enjoyed going to the occasional auction, if it was in the area. We'd often run into friends and neighbors and be able to catch up on what was happening in their lives. But most of the auctions were farther away, and Richard could never be sure when the horse equipment would be put up for sale. So, in addition to being away from the farm for work and travel, he left for auctions. I generally stayed home with Marian and Truman and was glad to do so. No sense in two of us standing around, too hot or too cold, waiting for a hay rake or a set of eveners to sell.

Late in the day, or even into the evening, I would hear his truck on the road, the old flat trailer he'd built behind it, a clank of metal as it passed over the potholes. We were usually outside and watched Daddy drive up with some kind of contraption chained and belted either in the bed of the truck or to the raggedy wood surface of the trailer. Sometimes both. He'd make a run around the back of the house to get the trailer positioned against the slight hill, making it easier to get the machine off, and then turn off the truck and emerge, stiff and tired from the day.

The kids were always delighted to see him. As for me, I must say it depended on what he'd dragged home. I was getting sick of the growing line of rusted equipment up by our new barn. Only some of it ended up being useful to us.

One time he hauled home an enormous threshing machine. He'd had to rent a larger trailer for the purpose and asked our friend Jay to help him. He was at the auction all day to buy it, then had to spend another day to go get it. When it arrived, I felt so annoyed at the waste of energy, time, and resources—I knew he would never get the thing running—that I hissed at him, out of Jay's earshot, "If you ever get that thing working, I swear to GOD I will set up a threesome with you, me, and one of my friends. I PROMISE you that."

To be sure, I had no intention of following through on that promise, but that's how confident I felt about the outcome of having that behemoth around—and I was right. Years later it went down the driveway on someone else's trailer, never having been used by us. And it was years after that before I reminded him of my vow.

***

There's a video I took one evening in that first spring after our first winter on the farm. Taking a video then meant holding the camera on your shoulder, it was that big. Just getting it out of its case was kind of a big deal, so I didn't use it a lot. On this particular evening you can see that the kids are impressed to be the subjects of the shoot—Marian rides her bike, training wheels bouncing on the gravel. Truman races on his small red tricycle, trying to keep up with her. They show me the tent we've set up. In the background, our cat Anna bounces out of the tall grass at the edge of what we've determined will be the lawn, which is spotty and rough.

We go up to the barn, and I film the harness area Richard has built into it with good, stout, white-oak boards. He's fitted the harness room with racks for harnesses, collars, and bridles. Above the harness room he built a loft where we can keep extra hay, and he's made a platform across some of the trusses where we can keep lighter items out of the way.

The corrals are built with pine logs—doomed, because in our moist climate they won't last three years before rotting. We didn't know that then, however. The railroad-tie posts, though, would last. I remember there were thirty holes to dig in the rocky clay, and even Richard quailed at the effort involved in doing that by hand.

We rented a hand-held auger for the task, racing home from the hardware store with it and rushing to get the holes bored. Richard pulled the starter cord and we grabbed the handles, ready to steady it as the auger ate down.

My arms were wrenched sideways—within the first six inches we hit a rock. This became the order of the hour (our goal was to get the device back to the hardware store within two hours, thereby saving ourselves about fifteen dollars). We'd make a few inches or even a foot, and then the auger would hit a rock, or bind in the clay, and we would be spun instead. We'd stagger in an arc around the prospective hole, Richard would do whatever one was supposed to do to make it—really, us—stop spinning, and we'd back it up, dig down with a spade or a posthole digger, find the rock, pull it up, and try again.

Arms, back, legs. Every muscle was wrenched. Once we'd gone deep enough, we hurried with the machine to the next mark we'd made on the ground and start again—usually with the same results.

We'd been on the place for one year. We had our wheat and potato crops in, we knew where we'd be cutting hay in a few weeks, and we had seventeen horses in the pasture. Stella was big sister now to Jimmy. Jemima had a colt named Jonah. Gidget was bred when we'd bought her with Martin at her side, and she'd foaled a little guy we named Gabriel. We'd bought a young mare named Belle, plus a young stallion, Ezra, at an auction in Indianapolis. Lucinda had her filly, Margaret. We still had Krista—an elderly Fjord mare—a Fjord stallion named Barney, and Inga. And Dinah and her foal, Rosalie.

Stella was two years old, our beloved pet. She pushed herself to the front of the herd to have her nose rubbed. Marian and Truman loved her best of the horses. One day Marian burst into tears when we were doing something outside among the herd. "Stella bit me!" she wept, more hurt that Stella would do something like that than from the injury itself. Examination of her little arm revealed not much evidence of the bite, so it must have been the lightest of nips, and we decided it was probably an accident. But good reason to be more careful around the horses in the future.

Teaching the kids to be easy around the horses, while being respectful and careful, was something we reinforced daily.

"You can't run faster than they do. So if they're running, which way do YOU run?"

"Sideways!" was the answer. I wanted them to run perpendicular to the motion of the herd. Sometimes the horses started running—a really beautiful and impressive sight, but not one that I wanted the kids to be part of—and swept up from the lower part of the pasture to arrive, panting and blowing and shaking their heads, at the gate.

They never did this when I needed them to do it, I promise you that.

"Should you stand behind something when the horses are hooked to it?"

"NO!" was the answer.

"Why not?"

"Horses can go backwards AND forwards!"

"Should you walk behind a horse?"

"Not unless it knows we're there!" And so on. The idea of what could happen to them, their small bodies meeting our horses' large muscles and hooves—it was a constant consideration, just as I suppose it is for parents who must raise their kids near busy roads or near water. They were older now, three and five, and they moved faster and were more daring in their adventures. Marian was starting to ride Krista, the old Fjord mare, surprising me with her authoritarian ways. I think Krista had not been ridden before, other than, of course, by me, testing her to be sure it would be safe to plop Marian onto her. She was willing to bear Marian's weight, but I think she felt as if that was plenty enough. Marian had to haul on the rein to get Krista to turn, and she used her heels vigorously to get Krista to walk ahead. I was always close by, but Marian was doing most of this by herself.

When Krista was ready to foal that spring, the kids and I arranged our sleeping bags and pillows on a hayrack in the barn for the night. I could check on Krista from my bed, simply flashing a light from where I lay. Luxury! But the night was very cold, and I carried the kids to the house in the morning—the grass was frosty, and I didn't want their shoes to get wet. Truman's cheeks were chapped red. He and Marian huddled in front of the heater as I made hot cocoa for them. No foal that night, but the next, when Richard was able to join us on the wagon, was more productive. She had a sweet little colt we named Harold, after one of my parents' friends—a kind, strong fellow, just as we hoped our Harold would be.

Richard and I built a paddock for our two stallions, Barney—the Fjord—and Ezra, the Suffolk. They got along surprisingly well, and it was convenient to have a place for the two of them to be together and apart from the mares. The day we finished the paddock, we let them into it during the day, so they would have time to get used to it before dark, putting hay down and leading them around the perimeter.

We were outside working till past dark and had just fallen into bed when we heard a commotion outside. Richard got up to check and came back with bad news. The stallions had gotten out and had run through the fences between them and the mares, and now all the horses were out. We had to find them, get them back into their respective paddocks, and re-build the fences. In the dark.

The light of the bathroom seemed to scour me as I dressed to go back out. I thought of my mom's admonition about farming and kind of wished I had taken it more to heart.

Chapter 8

We let the potatoes poke their crumply first leaves up before we first cultivated them. Later we would learn what a mistake that was. We didn't have a good way to cultivate the sides of the hills, either. The clean brown field became green, but mostly with weeds.

The wheat came up in a lovely quilt square of soft green. It was to be our first real wheat crop—the reason we were farming. We talked sometimes about the granary we'd have to build, and the milling room, and the equipment we'd buy. If I'd had time, I'm sure I'd have designed a label for our imaginary flour sacks. We liked to look at the wheat field.

Lucinda was going to foal that evening, as evidenced by her swollen udder and the wax icicles hanging from it. I was tired from working in the garden that day. I'd transplanted the tomato plants from our living room. Those were what I was worried about as the wind came up and the barn started to creak around me. I'd upended a five-gallon bucket over each plant, but with this wind the buckets would be blown away. I slithered out of my sleeping bag in the barn loft and went out into the dark to put a rock on each bucket—I'd picked several out of the garden that very day, in fact. I knew just where they were.

I was nearly finished with the job when suddenly the wind quit. The air filled with dust. I felt such a sense of danger that I lost interest in the garden and ran for the house. "Richard!" I called up the stairs, "Get the kids! Under the stairs!"

He appeared, blinking and slow. "Oh, it's fine," he said. Just then a window broke, the sound of glass hitting our floor galvanizing both of us to run for the kids.

We didn't have a basement. Our storm plan was to huddle in the space under the stairs, where the house was probably strongest and we could be farthest from windows and walls.

Zoom, zoom, the kids shot through the little door. I followed them, shoving aside toys and little boxes I'd been storing in the space. Richard was next, and we waited there for the storm to blow out, holding the kids and telling them it was all going to be just fine.

We learned the next day that it was straight-line winds, not a tornado. The wind blew the topper off the truck and rolled it into the hedgerow. It blew our plow sideways about six inches. It was hard to imagine a wind strong enough to do that—there isn't much for wind to grab on a plow.

Lucinda foaled early in the morning—a beautiful colt we named Atlas. The kids, once it was daytime and the storm seemed like an exciting dream, marveled at the broken window and were thrilled, astonished, and amazed to find a couple of earthworms on the opposite wall, blown all the way across the house.

***

I've never liked snakes. It didn't help, when I was a little girl, that my sister told me stories about snakes climbing up from the toilet bowl to bite people when they sat down to pee. Because I grew up farther north, we only had garter snakes, which are small, harmless, and a sign of health in the environment.

My parents had come down to see our new house, that first spring after we moved in. My mom and I were doing something when I overheard my dad and Richard commenting with mild interest on something outside on the bare dirt of what I hoped would one day be lawn. When I looked, I saw the most enormous snake I'd ever seen outside a zoo.

"Kill it!" My mother had no hesitation when she saw it. That thing was out there where her grandchildren played. Luckily, they were taking their naps at that time. It would not have been good for them to see how upset she was.

I felt sick—was this going to be a feature of our new home and farm? I didn't want it killed. But I wanted it gone.

"They eat rats and mice," my dad pointed out.

"So do cats!" I exclaimed. "And they do it for fun, not just when they're hungry!"

Finally, stirred into action by my mother's and my obvious disquiet, Richard got up and went out. The snake was just sunning on the lawn. It was a bull snake, about four feet long, completely harmless—yes, I know that. But still.

Richard used a manure fork to lift the snake, curling and writhing horribly (to me and my mom) and took it out of our line of sight. He brought it to the pasture fence, where the horses were gathered, resting from the effort of eating all that new spring grass. They looked interestedly at what he had on his fork. He held it out to them and then put it on the ground in front of them. One of them stuck out a nose to snuff at what this might be. The snake coiled and struck, shocking the horses, who wheeled and galloped away. The snake, Richard reported to us, laughing, slithered off in the opposite direction.

One afternoon Richard and I were baling hay. He was driving the tractor, I was stacking the bales. We were in a good rhythm. The old Ford baler was tying the knots reliably, and I was making a nice, tight stack on the wagon.

I picked up a bale and swung it up onto the stack, about even with my chest. I wasn't happy about how tightly it was seated, so I took it down again and this time hefted it with more force, getting momentum to lock it into place. I turned for the next bale and heard Richard yelling above the sound of the tractor engine and baler commotion. Finally I made it out. "Snake in the bale!"

I wasn't upset. I was feeling tough, I guess, and no little garter snake was going to keep me from getting that load stacked. I shrugged at him. He pointed at my feet. Not ten inches from my boots there was a severed tail—the last bit of a snake—a big one. This bit of tail, still twitching, was about six inches long. Where was the snake it had come from?

By that time I was ten yards from the wagon, trying to levitate above the hayfield. Richard took the tractor out of gear and walked back to the wagon. He took the bale I'd just stacked—and re-stacked—down and held the bottom out to me. There was an enormous (to me) snake caught in it, twisting to get loose.

I didn't swear a lot then—that came later—but I am sure I said something unprintable.

Richard took the bale from the wagon and set it on the hay stubble. He unfolded his pocket knife and cut the strings. The snake continued to struggle but could not escape. He pulled the flakes apart, making it less constricted, and he hopped up and kicked the tail piece off the wagon.

"That thing was hanging down when you were messing around with the bale," he told me. "And then you picked it up again! I figured you'd have a heart attack if it got loose and joined you on the rack."

Baling. The thing is, you really do have to make hay while the sun shines. I couldn't go have a shower and try to scrub the idea of the snake off and out of my mind—not till hours later. In the future, there would be a hornets' nest near where we stored the baler, and it seemed as if every time we went out, the first thing that happened was that a hornet would boil up out of the baler chute to hit and sting me. And the next thing was a bale of hay that had to be dealt with, and then another. Sweat and work had to be the remedy for the sting, as it had to be that earlier day for my shock about the snake. We had to finish the field, which we did. Not as quickly as we might have otherwise, though, because every bale that came out of the chute was suspect to me and I flipped it over and examined it before picking it up. Richard sighed and put the tractor into a lower gear.

The kids were with friends that afternoon. I went to pick them up after we'd gotten the hay into the barn. At dinner we told them about our snake adventure, and they were eager to go see if it had gotten away from the bale. We walked up the farm road, the kids hopping over the long shadows cast by the fence posts along the way, and we found the snake still in the bale—and another snake visiting it!

That is the truth, and I don't know what it means. In the morning the injured snake was gone, but I suspect a coyote got it.

***

One of the toys given to Marian when she was a baby was a Fisher-Price farm set. A cow, a pig, a chicken, a horse, a tractor, a little round-faced farmer, and a little red barn with a door that made a moo-ing sound when you opened it. No manure, no dust, no sour milk, no mud. No detritus from the winter's butchering, such as piles of cow entrails and the gnawed legs of the downed cattle, dragged onto the lawn by the farm family's dog—as had so often been the case at the farm where I grew up. I wonder at the misinformation Fisher-Price has caused in the minds of children, now grown up, who had one of these sets. Farms should be neat and tidy, the toy teaches, with animals who have no apparent purpose.

When I moved from central Minnesota to Illinois, I was amazed at the tidiness of the farms there. The neatly painted outbuildings were like ships anchored in the vast seas of corn and beans. It was only after some thinking about it that I realized that none of those farms had animals. The land was too valuable to use as pasture. It was to be cropped. And crops are neat, tidy things when grown with plenty of herbicides to keep the weeds down.

Our farm, new as it was, looked very different. The corral fences we were so proud of soon showed the effects of horses leaning on them, chewing on them, even running into them. In Fisher-Price-farm land, the grass grows right up to the barn. In real life, there was only mud or dust near the barn. In spring the horses churned the mud enough to kill the grass, and in summer they stood in the shade, switching their tails and stamping their feet against flies. They tamped the ground bare under the trees, and overhanging limbs were worn smooth with having been used to scratch itchy draft-horse backs.

Our pasture of beautiful, waving grass became cropped short as a golf green in places, and rank with weed growth in others. I've read that horses transfer the fertility of a pasture up to 40 percent because they dung in preferred areas and eat in others. Gradually over the years, our main pasture took on an irregular appearance—tall, weedy growth in one area, short clover-y forage in another. We could have prevented this through clipping it, but we had neither time nor equipment for the job. We used the American solution—more land. We made the pasture bigger and fenced new paddocks.

Neither of us had much farming or even gardening experience. We were too old, with too many other irons in the fire, to have become apprentices on a well-run organic farm where we could have learned how to do things the right way. Richard has a deep streak of the contrarian in him anyway—he preferred to "just do it." For better or worse, this became our unofficial motto. This is not the best way to go, though, and I don't recommend it.

We might also have advanced faster as farmers had I not insisted that we have a yard, flowerbeds, and even a picket fence around the lawn. I wanted the place to look nice, and I argued that prospective horse buyers and other customers would be in a better mood to spend money if their first impression of our farm was one of a pretty, cared-for yard. The farm women in the rural neighborhood of my childhood had taken pride in their yards, no matter whether they had to fit in the weeding of their flowerbeds after the evening milking or morning calf-pen cleaning. I understood their desire better than ever—I had it, too.

Richard was not inclined to agree, but for Mother's Day that year he gave me a pickup-load of rough-sawn cottonwood one-by-four boards cut in eight-foot lengths, plus a jigsaw. "Your fence!" he said with a flourish.

I staked out an area in the garage and set to work, cutting the boards in half, then using a large tomato-sauce can to draw a curved line at one end of each half. I used the jigsaw to cut along the curve. I soaked each board in a water-proofing product.

There was never a set time of day when I could work on the project—I just stole moments here and there. Truman enjoyed playing nearby, but Marian didn't like the sound of the saw. They no longer took naps, which had been hard for me to accept.

Richard was on another work trip when I put in the flowerbeds around the house. I dug out the clay that had been backfilled around the foundation, and I hauled that away in a wheelbarrow. I hitched the trailer to the pickup and drove into town to the local excavator we'd used when building the house. He loaded me up with black dirt, and I hauled that from the trailer, using the wheelbarrow, and dumped it into where the clay had been. You forget how heavy dirt is until you are hurrying around with wheelbarrows full of it, shoveling it out in quick bursts, knowing you only have so much time till you really HAVE to get back to work—to go catch a team and cultivate those potatoes, for instance.

My stack of pickets grew. Richard came home and helped with digging the fencepost holes. He put up cross pieces, and both of us of worked, when we could, to put up the pickets. I could see my vision becoming reality. Because there were no trees near the house, the building had a plunked-down look. The fence was to serve two purposes: to anchor our house on the farm, and to keep the farm away from the house. Inside the fence, I would rule.

I started the garden that year. Richard felt we should put it where the soil was best, and that probably would have been a good idea. But I wanted it near the house, so I could work on it as I could steal time from other chores. We agreed that the flat place to the south of the house would be most convenient and had fair soil. I knew I could improve the tilth and fertility of the ground, so that's where the garden went. I plowed it up with Katrina and Jemima, who I knew would not be too upset by the tight quarters, and disced it with them and our new-to-us five-foot disc.

We put in a small orchard to the east of the house and driveway, and we planted raspberry bushes between the house and the garden. The clothesline went opposite the raspberries. The asparagus paralleled the raspberries. I wanted to plant trees around the house, but Richard opposed me so strongly that I backed off on that one. He liked being able to sit outside on a summer evening with no mosquitoes bothering him. Trees would bring them near. I liked not needing screens on our windows until the June bugs came out, so I didn't put up too much of a fight.

Richard wanted something at an auction that was being held on a weekday, so the kids and I traveled to the sale and he went to work. I bought a reel mower—I remember how optimistic I felt about someday having grass to mow—and a pile of one-by-eight boards that looked good to me. Whether I was able to purchase the item he'd wanted . . . I don't remember.

I wanted the boards for a playhouse for the kids. Richard bought the two-by-fours necessary for the studding, and I called my dad to ask him to come down and help build the little house. He agreed, and he and Mom showed up early one Saturday morning with Dad's good hammer.

Richard and Dad had to call me to pull the kids off the job sometimes—they got too interested and were in danger of being hit on the backswing as a nail was pounded in, or bonked with a board being swung into position.

"Here's where my bed will go," my dad told Marian, gesturing along the side of the wee house. Marian was startled to think of that—Grandpa's bed would take up the whole space!

"No!" she exclaimed. "You can't live here. You have yours own house."

My dad enjoyed nothing more than having a good tease with his grandchildren. They went back and forth on it, Marian in great earnest, my dad seemingly completely serious.

My mother was enjoying it as much as I was, but she finally broke it up, coming in on Marian's side. "Grandpa, she's right. You have your own house. You don't need to live in hers. Let's go have some lemonade."

"Oh, all right," Dad said, all defeat and sorrow for a moment. And then, brightening, "But you'll have me over for dinner sometime, right?" Which, of course, was completely agreeable to his granddaughter.

***

There was an auction to be held about ten miles from us. That was luck in itself—auctions usually involved quite a drive. But this one was on a farm that had been the home of two bachelor farmers—hermits, really—and the auction bill listed a lot of antiques, which were not interesting to us, and farm equipment. Old farm equipment.

We were interested in the Allis-Chalmers All-Crop combine. We needed to harvest our wheat and had no way to do it. The All-Crop had been an innovation in its day—decades ago—and would be a big step forward from our scythe and bang-the-sheaf-head-in-the-children's-pool method of harvest.

The farm was just off the highway, and we parked in the long line of cars and pickups tilting into the ditch. Hay racks loaded with home goods were rolled out onto the rough lawn. My neighbor Leslie greeted me from her spot near a Red Wing crock she planned to bid on. "Go look at the other rack," she encouraged me. "Such cool old stuff."

The kids and I—Richard had gone straight to the equipment—wove through the crowd to the other rack. Old dishes, ornate vases heavy with gilt, dark frames holding black-and-white photographs of solemn men and women, an old carpet sweeper, twisted wire rug beaters, shoes that buttoned up the sides, a stack of moth-eaten coats.

Where had all this stuff come from? The woman beside me told me it had been in trunks in one of the sheds.

"The brothers moved out of the house and used it as a granary. They moved into the old chicken coop. You can see their beds in there." She rolled her eyes and made a face that indicated distaste. "I don't know what gets into people," she said. "Using the house for a granary!"

We moved on to the outbuildings. Sure enough, in one of them that was the size of a chicken coop, there were platforms nailed to each wall. Matted blankets had been kicked off and left at the foot of each; a gray, use-flattened pillow indicated the head. There was a small wood stove, and some clothes on nails.

Truman seemed much too interested in it, as if he thought this looked like a good way to live. I grabbed him and quickly moved on to find Richard. As I waited for space to clear in one congested area near the barn, I overheard a man telling another that the brothers had not believed in banks. "They just kept it here. When they went to buy a car one time, they brought a milk can full of bills!"

Whether that was true or not, it was clear that this place had been neglected for years. Somehow the two brothers had paid the taxes and kept themselves alive, but box elders had invaded what must once have been a planting of trees meant to break the wind and create a nice border between the farmstead and the road. The trees were snaggy, with dead limbs deforming young saplings. Cockleburs and nettles filled every open space.

We found Richard looking at the combines. There were two of them, identical. "I plan to bid on both of them," he told me. "I can use the extra one for parts."

As usual, it took most of the day to get around to auctioning off the old equipment. By then, the household goods had been sold, so the crowd was considerably diminished. Richard was able to get both combines easily. No one was interested in such old, out-of-date harvest equipment. Fortunately, for us.

We went back the next day to pick them up. An old, thin man in frayed clothing drove an IH Super M tractor out of the barn. He turned off the tractor and came over to us.

"Can I help you?" he asked.

"We came for the All-Crops," Richard said, and the man nodded.

"Okay," he said. "Let me go get 'em out of the bushes."

He started the tractor and wheeled off on it. "Is that one of the brothers?" I asked. "I thought they were dead."

"Me, too," Richard said. "But he sure fits the profile."

The All-Crop followed the rusty tractor back into the yard. As Richard fiddled with getting the combine hitched to our pickup for the ride home, the man stepped back.

"Is this your place?" I asked him.

He turned his gaze on me. "It is. For now. I'm Hollister Tharpe. My brother Clovis died a few months ago. I guess I'm moving, too. My niece found me a place that won't be so hard to keep up."

It didn't look as though this place had been kept up for a long time itself . . . but that wasn't what caught my attention.

"My husband was born in Clovis, California," I said. "And I've been to Hollister, too. Are you named for those towns?"

He shrugged. His shoulders were sharp as knitting needles under the thin fabric of his plaid shirt. "My mother liked the names, I guess. I don't know that she ever saw either place. But that's what she named us."

Every time I drive past that farm on my way to the next town, I think about him. The trees have been cut back, the old house razed and replaced, the sheds either knocked down or repaired. Someone else owns it, of course. The mystery remains—how did this seemingly normal, kind-looking man end up sleeping in the chicken coop on a bed that would not do for most people's dog?

Once we had the combines home, Richard set about learning to run them. Or, how to run the one. Not surprisingly, often, a part that was worn or badly rusted on one machine was in the same shape on the other. Nonetheless he replaced belts, found and pumped grease into the many zerks, mended what he could, and pounded dents out of the hopper. "Makes me wonder if this thing had a roll-over," he said about that.

He asked me to get him a wrench he'd forgotten in the barn, which I did. I figured he needed a drink right about then. On my way to the house, I passed Truman, who was slouched on his little red tricycle, legs spread, feet splayed, hands on the handlebars in Richard's big leather work gloves with their gauntlets—which on Truman went up to his elbows—and wearing a plastic astronaut helmet he'd inherited from his cousin Devon. He gave me a level, steady look and said gravely, "Watch out, Mom. Here come the naughty bikers."

Right, I snorted to myself, still snickering as I came back out with a fruit jar of water for my naughty biker's dad.

Before long, Richard had it running, mated to our Allis tractor and the power take-off necessary to turn the shafts.

***

One day I was raking hay with Ezra and Jemima. It was Ezra's first time working. We'd been using him singly to haul logs, and I'd hitched him with Jemima and driven him a few times. But today was the big day that he would learn the meaning of work.

Richard and the kids came up to check on us and to bring me a drink. "He's doing fine," I said, wiping my mouth with the back of my sun-warmed arm. "No problems." Richard was glad to hear that. We didn't want to keep a good, able-bodied horse around just for breeding (other than Dinah, of course). Ron and Deb had told us to give him a job, and we had plenty of those.

When we were finished with the field and heading home, Ezra broke into a little jog. The next thing I knew, he and Jemima were at a dead run.

My hauling on the lines had as much effect as if I were trying to pull a building down with my arms.

I was being tossed and thrown around on the forecart. The rake behind me was, luckily, out of gear. I braced my right foot under a bar of the forecart and hung on. We were going down the hill road beside the hay field. My desperate clench and pull on the left line began to have an effect. I was able to turn the team in a curve that led them to running up the slope rather than down it. They would tire more easily.

A wave of fury washed over me and I hollered at them, "Run! Run! Run!" I slapped them with the lines. I shouted and swore. Down in the yard, Richard looked up and saw, far off, what was happening. There was nothing he could do. In case the team curved around and continued to run down and into the yard, though, he wanted the kids to be safe. "You guys get up on the picnic table," he said. And then they watched, too.

Both Jemima and Ezra were tired when they started running, and now, having run up the slope, they were sagging. In spite of my urging, they slowed to a walk. Only then did I realize we had lost the rake—the joggling had loosened the linchpin. Richard showed up shortly after that in the pickup and we silently took in the damage to the rake—when it came loose, the hitch had burrowed a trough in the ground and turned back on itself, like a pretzel. A steel pretzel. Later, to fix it, Richard had to cut it off and weld on a new one. For now, I walked the horses back to the barn and spent a good bit of time cooling them down. And getting over the shakes that overcame me every now and then.

***

The wheat was drying back. The desiccating plants cast less shade, so there was more light for the understory of weeds, which quickly sprang up even taller than the wheat. The field went from looking pretty good to being a carnival of weeds—ragweed in particular. It was discouraging.

"Well, the screens in the combine are supposed to separate the wheat from the weeds," Richard told me when I worried out loud to him. "We'll just have to see."

We decided to give the combine the best possible chance by mowing the wheat first and letting it dry in the field. The weeds would dry, too, and possibly the weed seeds would not even go through the combine. I hitched the team and mowed one morning while the plants were still a bit damp from the dew. Our reasoning was that if we waited till everything was dry, we were likely to lose wheat berries themselves in the field, along with the weed seeds.

Two days later it was Saturday and time to try the new combine. We'd taken such a chance in mowing the wheat—if it had rained, our crop could have been pounded into the ground and lost. But for once the weather smiled on us. The tractor started, Richard hauled the combine up to the field, and after holding up crossed fingers at me and the kids, he put the tractor into gear, which started the combine's internal workings.

It clunked down the field, snuffing up the mowed wheat. Straw and chaff began to spit out the side, just as expected. Before making the turn, Richard stopped the tractor and combine and hopped down from the seat and up onto the side of the combine, peering into the hopper. He looked back at me, hesitating, then put up a tentative thumb.

Ok. It was kind of good. Better than not good. He came back down the field and stopped again. I ran out to look into the hopper for myself. I hoped to see it beginning to fill with clean, plump wheat berries, and certainly they were there. But so were wheat berries that were still encased in their husks. So were a lot of grasshoppers, alive and dead. So were various weed seeds. This was not wheat we could grind.

"I'll see if I can adjust it better," Richard said, and he got a set of wrenches from the toolbox on the tractor. He did what he could, following the faint instructions that remained here and there on the side of the machine. Most often he sought out the worn spots that showed where the lever or knob had most often been set—that was probably the way that worked best.

All afternoon he worked at harvesting the wheat. Up and down the field. Sometimes it clogged, and he had to crawl almost into it to dislodge the nest of wheat and weed stems that had bunched there. His allergies thundered up and he spent a good bit of time sneezing, wiping his eyes with his t-shirt, and—frighteningly—trying to get his breath.

The last pass was a huge relief. We'd emptied the hopper a couple of times into the old gravity box we'd bought at an auction, and now Richard augered into the gravity box the last of what was left in the combine. I climbed up the metal ladder on the gravity box and looked into it. Wheat, yes. And all the other stuff as well. Now what?

Well, there is a thing called a fanning mill, and we bought one. They've long been used to clean wheat. A modern combine is really a huge set of fanning mills, among other things. I'd read about Father Wilder cleaning the newly threshed wheat with a hand-cranked fanning mill in Laura Ingalls Wilder's Farmer Boy. Now I would be doing it. At least mine had an electric motor to run it.

This also was an auction purchase, and only the electric motor indicated it was newer than the one Mr. Wilder used in the book. One poured the wheat into the hopper at the top of the gizmo. There was a narrow opening

Holding the Lines

through which it would drop down onto a screen. The screen was tipped slightly to cause the wheat (once you plugged in the motor) to shuffle off the screen, past a fan, and onto another tipped screen. Every screen (or gang, as they are called) was a sheet of metal with rows and rows of holes punched in it. Our fanning mill's gangs had penciled handwriting on them in block letters, very much like my dad's handwriting, I always thought. WHEAT read one. RYE read another.

At the bottom, the cleaned grain collected and was shunted to one corner, where it fell onto wooden paddles that were attached to a roller chain, which raised the paddles and delivered the wheat to a chute that let it leave the fanning mill and fall into a burlap bag I could fit into place.

It was a clever contraption, and after understanding how it worked and how using the different gangs could separate weed seeds from the wheat berries at least part of the time, I could get a fairly clean bag of wheat out of it. Not clean enough to sell, and not clean enough to mill for sale to the public, unfortunately.

We'd taken a step, at least. We knew we had to have a less-weedy field, for one thing, and we had to improve the combine.

***

We couldn't sell the wheat . . . but there were the potatoes.

We couldn't possibly dig all those potatoes by hand. Richard spent another day in pursuit of a potato digger. This he found through the classified ads in a regional agricultural paper. It was across the state, and he got up early one Saturday morning to drive over, hauling the trailer, and bring it back.

The contraption looked a bit like a stegosaurus. The front of it—the head of the dinosaur—had a wide blade that came to a point. This would dig into the ground under the potatoes, which would be forced, with a lot of the dirt that was around them, up onto a series of linked bars—a chain, really, but with big, wide, flattened-out links—that were looped around uneven sprockets that turned when the machine was in gear and moving. The potatoes and dirt bounced up the front end of the stegosaurus, and in doing so most of the dirt sifted down through the chain and onto the ground below. The potatoes fell off the back onto another clanking chain, then onto the dirt that had formerly surrounded them.

The wheels of the digger reinforced the stegosaurus image. Every six inches, all the way around each wheel, there was a flange of metal sticking out about four inches, ensuring the wheels would grip when the machine was pulled forward, thus moving the chain and accomplishing the work.

I thought once again of how clever the people were who had figured out these machines. On the other hand, most of the time I spent in menial tasks I was thinking of ways to make them easier. Consider hundreds of years of that kind of thinking in rank upon rank of farmers—someone was bound to come up with something. And then someone would improve upon that. And so on.

It took four horses to pull the potato digger. Katrina, Jemima, Lucinda, and Jane were our go-to workers at that time. The first time I harnessed four horses abreast, before I went out to do it, I drew the configuration of the lines—outside lines to the outside bit ring of the outer horses. Inside lines to the outside bit ring of the inner horses. Bit-to-bit straps between the horses. I checked my work, imagining myself holding the lines, the horses restive in front of me. Should be okay, I decided.

The gravel driveway was hard on the digger's wheels. I drove them up on the grassy verge. Richard positioned himself over the start of one of the rows. We'd been unable to keep the field cultivated very well, and we had no way to continue to hill the potatoes as we should have throughout the summer. So it was tough work to see the actual row. We were often glad our farm was not near a road so people couldn't see how clueless and green we were.

I centered the row between the two inner mares. I put the blade down and engaged the gears. "Okay," I said, and the horses pricked up their ears, knowing I'd be clicking for them to go soon. Which I did, and they stepped forward. The row of potatoes, weeds and all, rose up and passed under me. It was remarkable. Because of the weeds, the potatoes didn't bounce cleanly off the back but were often tangled with weed roots, or still at least partially encased in dirt.

Richard stopped me and made an adjustment, as some potatoes were being cut in half. Another click and the blowing team surged into their collars, the earth becoming a thin river over which I made a tense bridge, holding the lines and bracing to keep my seat.

"It's hypnotic," Richard said when I stopped at the end of the row. "The way it all just unzips, and you end up with the potatoes on top of the dirt instead of underneath."

We dug just a couple of rows—we'd have to come back and pick up all these potatoes by hand, and that would take a while. We unhitched and left the digger beside the field, ready to go the next day, and we went back down to the barn—me to unharness the team and Richard to get boxes and bags to hold the spuds.

It was dirty, dusty work picking up the potatoes. The kids loved it—for a while. The blackcaps were abundant in the bushes beside the field, so they abandoned the potato endeavor and became berry pickers, adding smears of purple juice to the grime on their faces and hands.

"Well, we know now we can grow potatoes," Richard said as we flipped potatoes into burlap bags that we dragged along the ground until they became too heavy.

"And harvest them," I added. We could grow wheat, too. It was the harvest that was confounding us.

But could we sell them? A few days later I sorted out the ones that were dinged up or scabby and I rinsed the nice ones with the hose. I had a new, clean muck bucket, and I filled that and loaded it into the back of the minivan. I dressed the kids in better clothes than they normally wore on the farm, and I of course cleaned up my own look as well. We were going to the city.

I stopped at one of the larger co-ops and talked to the produce manager. He looked at what I had to offer. "We like to buy really clean vegetables," he said. "These are pretty dirty. And we have a potato source already. Thanks for coming by, though."

My first thought was, REALLY? You think THIS is dirty? He hadn't seen the potatoes lying on the ground, many of them still coated in our clayey soil. I was used to rubbing a bit of dirt off of even the ones I'd rinsed and brought into the house to cook for dinner. We were eating out of our garden that summer—beans and tomatoes and lettuce and spinach—and a bit of dirt here and there was just not a bother to me. I looked more closely at the heaps of potatoes and other vegetables on offer at this city co-op, though, and they looked like gemstones, they were so clean. Even shiny! There wasn't anything I could do to change the minds of those nicely dressed people roaming the aisles of the store in pursuit of their well-washed food. I'd have to step up my game.

I tried a few other co-ops and had similar luck. If only, I thought, I knew a produce manager at a co-op. And then I realized that I did! Aha!

Ron and Deb had hooked us up with their former interns, Barb and John. In fact, Barb and John had helped us assemble the other two sets of harness. Barb was a produce manager at Mississippi Market Co-op, which I now only had to find.

I had an idea of where to look and was able, by asking directions of a couple of women out for a walk, to get there. I brought Barb out to look at our potatoes. "They could be cleaner," she said, "but otherwise they look good. I'll take what you have."

Our first sale! It was a good day. I learned that we had to clean the potatoes better, and Barb gave me her number at the store to call every Monday to ask how many pounds they would need.

At home over the next few days I tried various systems of cleaning potatoes. I settled on a metal grate that had been the ramp for a trailer, and propped it up on blocks of wood. I spread the potatoes out on the grate and sprayed them with the hose nozzle screwed down to make the water spurt out with force. I used a push broom to roll the potatoes around so I could get at all sides.

That worked pretty well. It was slow, as I also air-dried them on the grate, but it wasn't as if we had tons of potatoes to wash at a time. Those days lay ahead.

Now I had a more attractive product to show produce managers, so I made more sales. A few rows at a time we worked our way across the potato field. We saw the growth habits of the different potatoes and determined which varieties seemed to thrive best in our conditions. We learned about an implement that hilled AND cultivated, and we determined to find one and buy it before next spring.

One of our neighbors was a professor at the local university. He liked riding his horse on our place, and we enjoyed chatting with him when our paths crossed. We were picking up potatoes one afternoon when he passed. Our conversation turned to the problem of getting the potatoes clean enough, and he said the university had a produce washer and he'd figure out how we could get in there to use it.

So it was that on some evenings we filled the bed of the pickup with boxes and crates of potatoes (we gave up on burlap bags—too hard to organize and harder to lift than the new plastic crates we'd purchased) and backed up to a door of the ag building at the college.

The produce washer was like an elongated clothes-dryer drum tilted on a frame. You dumped the potatoes in one end and started it, and it slowly revolved as jets of water shot into the drum. The potatoes caught on baffles, just as in a washer or dryer, and then slipped off as the drum made its slow spin. When they came out the other end, depending on how clean they'd been going in, they were clean. Or at least cleaner. We sorted them—this one was clean enough, this one needed another run through the drum.

The kids explored the lab with all its various machines for handling, cleaning, cutting, and packaging food. There was an area for hosing things, so we had Marian hose the dirty crates to be ready for the clean potatoes as they became ready. Truman was soon being hosed as well, but we expected that. They were happy, and that was the main thing.

Getting the potatoes loaded into the pickup for the drive to town, unloading them, loading them into the washer, sorting them into the clean crates, then loading them back into the truck, drying off the kids and dressing them in their pajamas for the ride home—they'd fall asleep and we'd just put them into their beds once we got there—then unloading the potatoes into the garage (to keep them safe from woodchucks and other animals that might gnaw at them ...) this took hours. Hours and hours. We had to find a better way.

Every season has a faint shadow of last year's same season—you compare your crop or pests or weeds to what happened last year. But mainly, you think ahead to next year's season. You do this so much when you are getting started. It's a kind of time travel. During those hours we spent picking up potatoes from the dirt or running them through the washer at the college food lab, we were thinking and talking about next year and how to make the work easier, more productive, more efficient.

We needed to find a way to wash the potatoes thoroughly on the farm. We needed a place to store them away from the threat of animals or freezing. And we needed more customers.

Chapter 9

Marian was five years old and ready to start kindergarten. She and I went to Kindergarten Round-up, saw the classroom, rode the school bus, met the teachers, and talked to other parents. The school did a good job, and she liked it. But she asked if she could go to school at home, and I loved the idea of having her around a bit longer. So we started homeschooling.

While we still lived in the split-level on the lake, but had begun talking about farming, I'd read a novella by Jane Smiley. It was in the short-lived magazine Wig-Wag, and it was called "Good Will." In the story, a couple lived out of the mainstream, off the grid, even off the road. They made, found, or grew what they needed. They lived in a way that I found compelling.

They had a son. The boy was old enough to be enrolled at the local public primary school. The story was about his actions juxtaposed against the actions and belief system of his parents. It was a good story for me to read at that time. What I took from it was the realization that my children would not necessarily love what I loved, believe what I believed, or want to do what I wanted to do. They would have their own ideas.

Why it took a fine short story to wake me up to this fact is unclear. I certainly was not living out my parents' dream for me. But because of that story, I resolved to let the kids participate as much or as little in the farm as they wished. They'd have to help out, of course. But we'd still do school, swimming lessons, sports, drama—whatever came along and they wanted to try. I hoped, though, that they'd want to join us on the farm. And that, I guess, is why I agreed to homeschooling my daughter, avid supporter of public schools that I am.

"School starts next week!" I told her one morning at breakfast. She was amazed, but excited.

"I can't WAIT to learn to read," she gloated, "a-cause you hardly ever read to us anymore."

That was too true. We had so many horses to feed through the winter that I was always, it seemed, mowing hay, raking hay, or baling hay. I was training the young horses—first fitting a harness on, then attaching a bridle, then, over days, getting them used to walking in harness, having something behind them, and being hitched with a quiet older horse such as Katrina. Once I had a horse reasonably adapted to the idea, I started using it to rake hay. The noise and commotion was frightening, but not so much to Katrina or whichever ho-hum mare or gelding was hitched next to the newbie. The green horse took its cue from the older horse, and it didn't take too many rounds before it was walking along, nodding its head, well on its way to being a good work horse.

Sometimes, when we didn't have to go to the college to use the potato washer, I mowed the lawn in the dark. I used the reel mower I'd gotten at the auction. It didn't wake the sleeping children, for one, and it made a good cut. I liked the quiet, and the lack of gasoline smell. At any rate, Marian was right: she and Truman were being put to bed with the briefest of stories. No wonder she wanted to learn to read.

We developed a routine. Homeschool in the morning. When Richard came home from work, I had the horses harnessed and ready, and we all trooped up to the potato field. I dug a few rows, then brought the horses back and unharnessed them as Richard and the kids started picking up potatoes. I hurried back up to start filling crates, too. We all watched for the perfect cull—a beautiful potato that had been dinged in some way, usually by the potato digger. It couldn't be sold, but we'd have it for dinner. We placed our culls in a nest Marian made in the bed of the pickup; after a few of them were settled in there, it looked as if some odd bird had laid a clutch of beautiful spuds and they'd each taken a bite of each other.

Usually, the next day I spread the potatoes out on the grate and did a preliminary wash with broom and hose. Having them fairly clean going into the produce washer at the college made our evening potato-wash marathons go a little faster.

Once a week I delivered potatoes to Mississippi Market and the other co-ops that purchased from us. Starting with our small local co-op, of course, I'd follow my list from River Falls to Hudson to St. Paul. I tried to synchronize it with the kids' down time—I could usually count on Truman falling asleep in his carseat, his little blond head drooping to the side and his cowboy boots sticking out. Marian and I enjoyed our conversations.

"Mom," she said one day. "I am finking. I am always finking about somefing." She paused to look out the window. "I am finking right now!"

Another day, she seemed troubled. "Mom," she said. "What if you have a baby in yours tummy and you don't want it to be there?" I still remember the little exhalation I made upon hearing that. Where to start? How much to tell?

When our cat Roy disappeared after the Halloween blizzard, I explained that he had died. Because she was going to Sunday School at the time, she was aware of this thing called heaven. Did Roy go to heaven, she wanted to know.

"Well, the part of him that thinks went to heaven," I said—meaning, I guess, his consciousness. That's the part of a person I think would be most likely to live on after death, if that were possible.

Some months later she asked, "Mom, up in heaven. What do they do with all those HEADS?"

Truman was learning to talk. He spoke as if he'd been programmed, but badly. "Mom." He used a period after every word, and he always prefaced his statements with 'Mom' and waited for me to respond.

"Mom."

"What?" I had a "new" (garage sale) sweat suit on.

"Why. You. Look. So. Burful. Today."

Oh, so he thought I looked beautiful today in my new clothes! Another reason to smooch him. There were always plenty of those.

***

It was late autumn, and the chickens were scratching for bugs in the drying grass. Marian and I were waiting for Truman to put away his tricycle so we could go to the woods to pick colored leaves for a poster she was making. The chickens scratched and pecked, scratched and pecked, almost right up to our feet.

Marian pursed her little lips in a kind of annoyance. "These chickens have no skills," she said.

A few days later we were checking the fences, and a V of geese flew over, catching our attention by honking and honking. "They are heading south!" Marian exclaimed.

"Yes," I agreed. "They know where south is."

"Now THEY have skills," Marian said, approvingly.

Through the summer and fall, when he wasn't traveling, at the office, or farming, Richard built a loft the entire length of the north side of the barn. Below the most eastern section he set up a bench and shop for himself. Next to that, he built in a foaling stall, sixteen-by-sixteen. Our mares would foal out of the wet and cold if we were vigilant enough to bring them in for the event.

Next to that he built an insulated room in which we would store our potatoes. This meant we could plant more, harvest more, and sell them longer into the winter. We stored our small crop in there, imagining the day when it would be filled.

Holding the Lines

***

For all the counting back we did—all the figuring and reckoning—we could not figure out how and when Dinah got pregnant. That time the stallions got out in the night? That had to have been it. Dinah's udder was swelling, and Jemima had a big belly on her. Winter foals were a bad idea. We never wanted a foal born in winter, when it could so easily freeze.

But here was Dinah, waxing on a day when the temperature was well below zero. And the temperature that night was to fall to between thirty and forty degrees below zero. One of us would have to stay in the barn with her. Richard grew up in California. Decades in Minnesota and Wisconsin had hardened him, but his nose still froze easily, and he simply avoided, when he could, cold weather. I was the better-insulated one, I liked being with foaling mares, and I was used to the cold. We agreed that I was the one who should sleep in the barn that night.

So it was that I went out into the darkness with a hot-water bottle and two sleeping bags, dressed as if to hike in Antarctica. I turned on the light over Richard's workbench so I'd be able to see Dinah in the dimness of her foaling stall. I patted her nose and looked her over, taped her tail loosely, and then fitted one of the sleeping bags into the other and worked them into a recess I'd made in the hay. I slid the water bottle down to the foot of the bags and slid myself in after. It wasn't so bad, especially after the cats found me and curled around my head and neck.

I was good at falling asleep in the barn by now. Before long I was waking up—the sound of water falling. Dinah was peeing. Back to sleep. Then awake again—this time Dinah was up to something. I wriggled out of my warm slot and saw her back arched, steam rising from her sides. I ran to the house and called up the stairs for Richard, who answered quickly that he'd be right there.

Then back to the barn, where Dinah was lying down, then getting up. Richard arrived with the bag we'd packed full of towels, an electric hair dryer, and the Nolvasan I now used to treat the navel.

The foal's hooves protruded, and we watched it being born, leaving Dinah to handle it on her own. Once the foal was mostly out, though, we slipped into the stall and began drying it. As the towels became wet, we slung them over the stall gate, where they froze instantly stiff. Richard plugged the hair dryer into the cord we'd stretched out in preparation earlier that day and blew warm air onto the foal. Dinah was great, lying still and letting us work, keeping the umbilical cord intact so the foal could get the benefit of her warm blood a bit longer.

It was a filly, we saw, as we worked over her. She had a dramatic blaze—

as clean and white as it would ever be again in her life. She grew encouragingly impatient with our ministrations and wanted to get up. We stepped back—Richard ready with the hair dryer to get any exposed wet fur once she was up—and let her wobble and plop and work her way up onto her stout but shaky legs.

From there it wasn't far to Dinah's udder and a good, long, noisy suck. Richard and I had almost forgotten how cold it was. The relief of seeing Dinah and her foal warmed us momentarily, at least. But before long our faces were burning with cold, and we decided to leave mare and foal alone for a couple of hours. In bed, we held each other till we'd warmed ourselves and the sheets. We smelled of foaling—a scent as distinctive and clinging, I have found, as chlorine from a swimming pool. In the little warm pod we made, we held between us our hope for that filly, so new out there in the freezing barn.

She was fine in the morning. Full of milk, beginning to frisk. The weather remained barbarously cold. I sewed a blanket for her, using the materials I used to make warm hats and mittens for the kids. A red nylon outer shell, a blue fleece lining, two layers of quilt batting as insulation, and gold corduroy and velcro for the straps to keep it on her. The low that night had been 38 degrees below zero, but we decided not to be too hung up on exactitude and named the baby Thirty-Below Elaine. In her new blanket she looked adorably silly—an unlikely superhero.

A month later, in warmer weather, Jemima foaled without drama. Her colt was small and squarely built. Marian named him Blue, which she said was a good name for a winter baby.

***

"Good morning, Teacher," Marian said shyly as she came into the room we'd arranged into a sort of classroom for her homeschool.

"Good morning," I greeted her. "What is your name?"

"It's Mary," she said as she took her seat.

It was becoming more evident that she needed more from school than what she was getting at home. Every day now she took on a new name—Mary, Mandy, Elizabeth. She was restless. Even so, she claimed to want to go to school at home.

Our friends Rob and Nanette were sending their son to a Montessori school in the neighboring town. We decided to visit, and when Marian saw the activities and other children, she was hooked. Nanette and I shared the driving for the three days each week that our kids went to the school, and Marian thrived in the new environment.

On February 2nd, I explained the idea of Groundhog Day to her. She was fascinated and wanted to watch the news with us that evening. It wasn't on the news that night, but the next day there was an article about Puxatawney Phil, the groundhog, and apparently he did see his shadow. I showed it to Marian and told her what it said. She was terribly disappointed, and in the following weeks, whenever it was cold and gray, she would gripe, "Oh, if ONLY that groundhog didn't see his shadow!"

Prescott and Juliet, Nanette and Rob, and others we knew from our involvement in the food co-op started talking about starting a Montessori school in River Falls. To my surprise, Richard was enthusiastic about it. Our little group began to gain traction on the idea through the winter, and a few parents became deeply committed to the project. The idea began to take shape as spring came upon us.

***

Back when we had been imagining the farm we hoped to find, it had included a sugar bush. We wanted to make maple syrup as part of our farm income. Alas, our 200 acres had not even one sugar maple on it. But when life hands you lemons, so the saying goes—we had a lot of box elders, which are in the maple family and do produce a sugary sap. Just not much of it.

Nonetheless, we were determined to just do it. After some reading up on the process, equipment, and techniques, one Sunday afternoon we all trooped out to the fence line around our south forty, where the box elders had been growing enthusiastically for decades.

Richard used a bit and brace to bite into the first tree. I set the spiles and hammered them in with light, firm taps. Then I hung a sap bucket to catch whatever might bleed out. By that time Richard was finished with the hole in the next tree, and so it went along the line. The kids watched with interest and took on the job of putting the spiles into the holes, waiting for me to tap them in, and then hanging the buckets. The ground was still cold, and in only a couple of cases was there a drip of sap coming from the tree, but we held out hope.

The next morning, Truman and I went back up the hill, punching through the icy snow in places where we could not follow yesterday's footprints, and found . . . nothing. Later that day, after Marian was home from school, we tried again. This time we found some spiles dripping—even, in a few cases, an accumulation of sap in the bottom of the bucket.

As the week wore on, the weather became what one hopes for when making maple syrup. The nights were cold and the days were sunny. The trees seemed to get the idea, and we began finding buckets half and even three-quarters filled.

Richard had been building another protable horse shelter—just a three-sided shed—up by the house, and we set up a woodstove in there and bought a flat-bottomed steel pan that was a foot wide and two feet long. I could get a fire started and heat the sap to boiling and go back to whatever other work I was doing.

I was taking Marian to school one day when she said that she had thought "tap" meant something else. "Oh," I said, and I reached back and tapped her leg. "You mean 'tap' like that."

"Yes," she said. "But this other kind of tapping is WAY MORE FUN!"

***

Our beautiful Katrina had lost her first foal, Lucky Romper. Her second foal was born the following spring, a beautiful colt we named Carl. He was healthy and perfect. Now she was pregnant again, nearing her foaling date, her udder filling as expected. I readied the foaling stall, shaking out several bales of clean straw to make a nice, deep bed for her. I was extravagant with straw in the foaling stall mainly because I wanted it to be clean and comfortable in there, but also because after the event, I mulched the raspberries with the used bedding.

Katrina waxed. I wrapped her tail, which I'd learned to do to keep the hairs from entering her vagina as the foal moved out and then back in before emerging again. Whenever I did it, I remembered my mom, a nurse, telling me how hair could not be made sterile. As darkness approached, I put her into the foaling stall and hoped we would have the same success as the previous year. A filly would be nice, but another Carl would be fine, too.

I checked throughout the evening. When I went out to slip into my sleeping bag to stay out there for the night, I found her down, the foal halfway out, a piece of placenta draped over its head. Just as with Lucky Romper, the placenta had detached too early. Katrina pushed it the rest of the way out and lay quietly, catching her breath. The foal panted. Well, of course it would pant. It had come a long way. All foals pant.

But it kept panting. It wanted to get up and struggled to do so. By this time Richard had joined me, and we finally helped the little colt to get up, stagger around, and finally, after an hour had gone by, find the nipple and suck. We were relieved then and felt as if he would be okay. We went in and slept the few hours we had till dawn.

I felt a cautious optimism as I approached the barn. It slipped away when I saw the baby. He was awake and alert, but laboring for breath. Katrina stood over him, and I felt her worry in my bones.

Peter came and diagnosed Respiratory Distress Syndrome. He recommended cuppage, explaining that four times daily I should beat on the foal's back with my cupped hands, trying to dislodge whatever was clogging his lungs. "Keep him up on his sternum. It's easier for him to breathe that way," he said. "It will probably be a week before you can relax."

Our foaling journal from that birth reads, *Now it is 3:15 AM. I've been up with the foal and his signs have not improved. He is weaker than ever. I got him up at 9:30 to nurse, then at 12:30, but at 2:30 he just couldn't get up. He is a beautiful and dear little thing, perfectly formed and so soft. Truman wants to name him Superman. I want so badly for him to live. Katrina is being so noble, standing over him and doing all she can to assist him in drinking.*

In the morning Richard had to go to work. We agreed that I would call the University of Minnesota and see about taking the foal there for help. When they said they could take him, I called Earl, our horse hauler, and he came. We carried Superman (we let Truman name him) into the trailer and Katrina loaded easily after him.

At the veterinary hospital they put the foal on a gurney and kept him under Katrina's nose as they worked on him. She stood over him quietly as students and vets busied themselves with shaving her baby's coat to allow for needle sticks and IVs, as they palpated and listened and looked into his mouth and eyes. I stayed as near her as I could, hoping my presence comforted her—as hers did me. She was the largest horse in the hospital and, to my eyes, the most beautiful—the most noble, though it may seem odd to say that about a horse.

In the end, there was nothing they could do for Superman. They euthanized him, also under Katrina's nose. I called Richard, who heard in my voice probably more than I wanted to let on, and he left work to be with us. We loaded Katrina in Earl's trailer—poor Earl, almost as sad as we were, dear man.

I called the vet hospital a day or so later to see what the autopsy had revealed. The young man with whom I spoke said, "It was strange. His lungs weren't developed. Usually when a lung has been oxygenated, you take a bit of it and it floats in water. But this foal's lung sample sank like gravel."

Horses are not people, and though you love them, losing a horse is not the same as losing a parent or a child. Even so, as anyone knows who has loved and lost an animal, you grieve. I moved around my house, cleaning it as if I could change something by having a shiny floor, and as I did so I told myself a story that has stayed with me through the years.

Superman did not die. When they put the fluid into his veins to euthanize him, it cured him. He stopped panting and breathed normally. He struggled to get off the gurney, and the astonished people around him helped him down. Katrina nuzzled his back and hip, guiding him to her udder, and she let her milk down in sheets.

He grew up and was beautiful. He became our herd stallion and loved being near his fillies and colts. He was easy to train and worked steadily in our fields.

One day, when he was very old, he lay down and he died, painlessly. We allowed the veterinary college to take a sample of his blood and to biopsy his lungs. The little piece of his lung, once it left his body, rose and floated in the air. They took it and made a serum from it and now no foals die of what did, in fact, cause Superman's death.

A grown woman kept vacuuming her living room that day, unwilling to let her kids hear her crying. You do what you have to do—as many women know who have suffered much more important and devastating losses than I have.

***

The man who'd purchased Jimmy and Jonah as weanling colts had decided, now that they were nearly three years old, that he didn't want them after all. He asked if we'd buy them back, and we agreed. They were geldings now and had been started in harness. Our intent was to sell them to someone else, but first they needed to be trained.

It was a clear, bright, sunny day. Just a little wind. Richard was in Europe. I was outside ground-driving the geldings, pulling a chain around and getting them ready to be hitched to a stone boat, when a car pulled up. It was kind of amazing to me how people could find us, as far from the road as we were, but in fact all they had to do was get into the neighborhood and someone would direct them.

The older couple, Rudy and Esther, wanted to see the horses. Rudy wanted to buy a pregnant mare so his grandchildren could experience a new foal and see it grow. He was spry, with lively blue eyes. Esther was a perfect grandmother type, kindly and soft. As Rudy greeted Jimmy and Jonah, taking each horse's head in his hands and scratching in the right places, I watched Esther watching him, taking obvious pleasure in his pleasure at being with horses.

We'd decided to sell Jane, and I'd placed an ad in the regional agricultural paper. I'd trained Stella a few months earlier—an easy task, as it turned out. She was oblivious to the harness. I drove her a bit from the ground, then hooked her with Jane. Once she seemed comfortable, I stepped Jane over the pole of a forecart and hitched them up. Stella was clumsy, banging into the pole, stopping and then being surprised when the evener ran up behind her. But she figured things out as she went. I was starting to use Stella with Jemima now more than with Jane, and we thought, given Jane's age, that she'd be happier in a place where she'd get more attention and less work. Someplace like Rudy's promised to be.

Rudy and I agreed on a price and that I'd have Jane delivered once his check had cleared. He lingered amongst the herd—the horses were beside themselves with their joy at having someone else to give them attention. The sun was weak but nourishing all the same—we were so hungry for springtime, all of us. Rudy and Esther drove away, delighted themselves and leaving the rest of us a little bit happier as well.

That night, before I went to bed, I checked on the kids. From Truman's window I could see the northern lights, a display so vivid that I felt as if I were on the bridge of the starship Enterprise, headed for a beautiful far-off galaxy at warp speed. It was too much for me to see alone—I felt I had to share it with Richard. I tried calling him in Belgium at the number he'd given me, but I couldn't get through. I stood at the window with the portable phone in my hand, boldly going nowhere for a while until the lights dimmed and I went to bed.

The next day I looked into the sky as if I might find some evidence of those colors and the slow-waving, majestic curtains. But of course there was none.

Chapter 10

We were gaining experience in using our horses and equipment, and we were a couple of years into our rotation. Our rocky, clay-ey soil was beginning to seem arable, what with the minerals we'd added and the carefully considered cover crops we'd plowed into it. The wheat crop went in and would nurse the alfalfa beneath it. We planted barley to see if we could grow it, interseeding that also with a legume.

The new potato planter was the best new-to-us thing we had, in my opinion. Richard bought it at an auction and brought it home on the trailer in triumph. "Now all you have to do is drive the horses," he said.

It was a hopper on wheels. I sat on a seat that was bolted to a metal strip that rose from the pole. There was a seat behind the hopper. Richard sat on that, monitoring the roller chain that was fitted with metal cups every six or eight inches. The roller chain passed through the hopper, where the metal cup would catch a potato seed piece and deliver it to the furrow that had been opened with discs that ran ahead of the hopper. Discs below Richard's seat closed the dirt into a pursed hill over the seed pieces.

If every metal cup was filled with a seed piece, the potatoes would be evenly planted at the correct distance apart. But often a cup failed to capture a seed piece. That's why there was a second seat, where someone, usually Richard, sat with a bucket of seed pieces, ready and quick to fill a passing empty cup.

It had to be refurbished, which Richard did very well, knowing how useful this implement was going to be for us. He made it look sharp—he ground the rust off the hopper sides and painted it green, put a stout wooden pole on it and painted that white, and fitted it with nice, sturdy, deep implement seats so we'd feel more secure as the planter rocked below us.

We put in more potatoes than ever. The planter required that I keep the horses at a slow, slow walk—otherwise the metal cups passed too quickly to be filled. This was a challenge. Some horses were simply not useful for this. Jemima could be held back, and Stella, though newly trained and heavily pregnant, was another good pick for the work.

Unless I couldn't keep the horses from walking too fast, which irritated him and frustrated me, Richard and I enjoyed this new method of planting. It was companionable work to fill the hopper, flip the seat down, and start off down the row. The horses could do the work easily, though on warm days they broke a sweat, and that scent would mix with the late-spring smells of warm soil and sun on new grass.

***

One evening I came home from book club and, before going into our dark house full of my sleeping family, took a flashlight and went up to the barn. Stella was in the foaling stall, udder swollen and possibly waxing—it had been hard to tell that afternoon when I'd put her in there.

I could walk up to the barn by moonlight, but once inside I switched on the flash and aimed it at the straw, hoping not to disturb Stella more than necessary. Her rear end was toward me, and her head cranked around to face me. I saw the sac with its small package of hooves beginning to emerge from under her tail. I switched on a less-intrusive light I'd clipped to a joist, turned off the flashlight, and sat down on a bale to watch Stella have her filly. We named her Baldur's Tribute to Jane, in honor of her grandmother. Baldur was our prefix and the name of our farm, chosen in a spate of enthusiasm for Nordic mythology.

Stella proved to be an excellent mother. Janie grew up to be what we'd hoped—a tribute to Jane. That night, I was happy to have arrived for the birth; after tending to the navel and being sure the filly got up and sucked, though, I enjoyed spending the rest of the night in my own bed.

***

On a lovely May morning, Wisconsin must be the most beautiful place in the world, or so I often tell myself. On this particular morning, I was driving back from Durand, where I had made an important purchase. The next day was Truman's birthday and we'd decided on the perfect gift for him. Our friends Lynn and Jay had one for their girls—a heavy-duty tricycle with a hitch. I'd checked around, which led me to Durand—about thirty miles away—and this day's happy drive. The mood of the whole John Deere dealership lightened when the guy behind the counter rolled the shiny green trike with its big, knobby tires, yellow seat, and "10 years or 10,000 miles" guarantee out to me. Everyone had a small, secret smile, remembering no doubt their own experience with a small toughie who liked tractors and hauling things.

I hefted it into the bed of the pickup but decided it would roll around too much, so I wedged it into the passenger side of the cab and started home. As I mentioned, it was a beautiful morning, my favorite month, and I had this great present for my funny little son's fourth birthday beside me.

And then I heard the siren, saw the lights, and noted that I was well over the speed limit for country roads in Wisconsin. The air floofed out of my mood balloon and I pulled over. The deputy took some time about getting out of his car and walking up to my window, where I handed him my license and registration.

We had the expected exchange about my driving habits, and then he noticed the trike beside me. "My son turns four tomorrow," I explained. "I had to come over to Durand to get it."

"Fun ahead," he said. "You drive careful now. Watch for deer." He rapped the top of the truck cab his knuckle and walked away.

And that was that for my speeding ticket.

As for the birthday present, few gifts have been more of a hit. He already had a wagon to match it—a wagon with removable sides and two interchangeable handles, one with a handle to grab and pull, the other designed to be hitched with a pin. We had the wagon set up with the latter and he immediately hitched up and began his career (for the next few years, at least) as a guy who pulls a wagon full of random things with his monster trike.

The day of Truman's birthday party, his friend Phillip's mother, Kim, looked out the window into the pasture. "Is that a new foal?" she asked.

I looked and sure enough, it was, way down in the pasture near the woods. My dad and I went out to bring mare and foal up to the nursery paddock where the little one could get used to being with its mother without the distractions of the larger herd.

As we approached, I could see that something was off. The foal was, by that time, lying down, fast asleep. Katrina stood over it, for all the world as if it were her foal. Which, of course, it could not be. She'd just had and lost her baby, Superman.

Gidget was the pregnant one nearest her due date. She was pacing several yards away. Low in the pecking order, she'd given birth and then been chased off by the superior Katrina.

I explained the situation to my dad. I put the halter on Gidget and we began leading her, with her wobbly baby beside her, up to the barn. Katrina kept trying to cut the foal away from us, scaring Gidget and confusing the foal.

"Holy Arlen," my dad panted. "She really wants this baby."

Once we had Gidget and the colt—soon to be named Spike by birthday-boy Truman—safely in the nursery paddock, Gidget could relax and let the foal nurse. She stayed well into the paddock, far from the fence shared with the larger pasture. Katrina stood at that fence for a couple of days, oriented like a compass needle at the baby she wanted to be hers.

***

We'd found a potato hiller and brought it home to the farm. All it needed was a new pole, which Richard mounted on it, and I hitched a team and went to work in the potato field five days after we'd planted the spuds. The potatoes hadn't sprouted, but weeds were beginning to green the sides of the low hills. They were tiny and vulnerable at that stage, and I meant to kill them.

The hiller was a set of wheels joined by an axle. I sat on a seat between the wheels, above the axle, and placed my feet in metal stirrups that could move the discs and sweeps, which were attached to the frame of the implement, back and forth as necessary. At the head of the row, I positioned the horses, one on each side of it, and used the lever to lower the discs an inch or so into the ground. Then a click to my team and off we went.

Suffolk horses tend to have a fast walk, which is great for cultivating. Throwing the soil up over the small weeds is part of the strategy. The discs were tipped in, like the ones on the potato planter, but in this case they were larger and more aggressive.

I checked behind me—the row was higher, and the sides of the hill were no longer greenish. Hours later, when I'd finished, the field again looked like magnified corduroy. What a great machine! In later years, I devised modifications to it—my most successful was a set of three leaf rakes. I saw that sometimes the hiller discs didn't get close enough to the hill to interfere with the tiny weeds. So I mounted the rakes to scratch along the sides and top of the row. As long as I didn't have to back the team, or turn them sharply, we were fine. I snapped a couple of handles before I perfected a technique that worked with the added hardware.

I hilled the potatoes every five or six days after that, weather permitting. They were well up and leafed out, nearly ready to blossom, when I was doing probably the final cultivation. I always damaged a plant or two when I hilled them—sometimes a horse stepped on a plant by mistake, or we went off track, or a plant was too far to one side and got cut off. But as they grew larger, it became easier to do real damage to more than a plant or two. The plants sprawled over the hills, shading out the weeds, and we counted on that.

This time I noticed potato bugs. I have to hand it to those evil little insects. This land hadn't had a potato on it for years and years. For twenty years at least it was just grass. But somehow they found us after just three years—our third crop of potatoes.

When I got back down to the house I called Richard at work—it was long distance, so I didn't do it often—and reported the bad news. Colorado potato beetles can devastate a crop very quickly if they get the chance. "We'll have to go pick them off," he said. "There's nothing else we can do right now."

I'd been expecting that response, but I had HOPED for a different one. The kids were with our neighbor for the morning; once I had them home and full of lunch, I explained how we were going to spend the afternoon.

It was interesting to them at first, but picking the bugs off and dropping them into a jar of soapy water got old in a hurry. I hadn't expected anything else. I made them each fill a soup can with bugs before they were allowed to wander off to find more interesting insects and to build a fort in the tall grass at the edge of the woods.

When Richard came home, he joined me. We were lucky—the beetles had not got too far ahead of us. We checked under the leaves for their bright-orange eggs and squashed those. Their younger larvae were still small and not as disgusting as the fat, orange, dotted blobs they would become very soon.

We'd been expecting bugs, at the same time hoping our remote location might delay them. Committed as we were to being organic farmers and not using synthetic chemicals on our land, we had limited options to defend our crop. The most immediate was just picking the bugs off, as we were doing. Another was Bacillus thuringiensis, Bt as it is most commonly called. That's a bacterium that grows in soil and which, when ingested by beetle larvae, through a series of digestive manipulations that take place in the larvae's gut, causes the little beast to stop eating and starve to death.

You would really have to spend some time in a hot potato field, picking those voracious orange larvae off your plants, to take as much delight as I did in the idea of whole swathes of them starving to death.

Eventually we purchased some Bt, first checking to make sure it was approved by the organic certifying agency that we were working with, and mixed it up with water and I used a backpack sprayer to apply it. Later that day the kids and I went up to survey the situation and found the bugs to still be on the plants, but no longer chomping away. We continued to pick the adults, and there were always larvae that escaped my spray efforts. For the most part, however—that season, at least—our potato plants continued to grow in relative health.

Holding the Lines

***

Richard bought a hay loader from a retired farmer in Red Wing, a small city across the river in Minnesota. It was in wonderful condition—this was so rare an occurrence, he was truly delighted to bring it home and not have to fix it up. For him, our farming adventure was involving a lot more time using tools and banging his head on the undersides of big equipment than he'd ever anticipated.

We weren't sure how to use the thing and asked our friends John and Barb, the former interns for Ron and Deb, to come out and make hay with us one Sunday. I'd mowed and raked the high field, which was our name for the field that was on the highest point on the farm—a windswept place that brought to my mind Tennyson's lines, *Close to the sun in lonely lands.*

I hitched a team—I have forgotten which horses—to the forecart and backed up to a hayrack—something I never got good at, I must admit—and Richard put the pin through the holes, hitching the rack to the forecart. The loader was already up on the high field.

We all made our ways up there. I pulled in front of the hayloader and hoped I wouldn't have to try backing the hayrack up to it. Luckily, John, Barb, and Richard were able to muscle the loader close enough to hitch it to the hayrack. So now it was the team, the forecart, the hayrack, and the hayloader.

I clicked to the horses, and we pulled forward over a row of raked, dry hay. Oops! There was no mechanical noise behind me—I stopped. Richard jumped forward from where he was watching and put the loader into gear. I started again—the horses were startled at the noise and resistance but soon settled into their work. The whole parade moved along the windrow; the hay remained on the ground, and then, as if an afterthought, the tines of the hayloader grabbed it up and sent it in a kind of reverse waterfall up the metal slope, cascading down onto the hayrack.

"Wow!" Richard shouted to my questioning look. "It's working!" He hopped up with John onto the wagon, each of them armed with a three-tined fork, and they began distributing the hay to the sides and corners of the rack. Every time I rested the horses, I looked back and saw a higher stack. John and Richard used the resting time to rest themselves—it took a lot of stamina to stay balanced on the spongy load while moving big forkloads of hay as fast as possible. Barb used rest time to jump up and run back and forth on the hay, pressing it down so we could fit more onto the rack.

"I should have hitched another horse," I called back. "It's a tough pull for two. With three we wouldn't have to rest so much."

"We need the rest!" Richard and John together shouted back at me.

Once the load was huge and toppling, they unhitched the loader and hopped up to ride on the hay as I drove back to the barn, where we forked it into the corner of the barn. In times gone by, many barns were set up with large hay forks or grabbers that came down and either plunged into the load or grabbed at it, while the team was unhitched and waited outside for the signal to step up. A rope was slid over a pulley, and the horses jerked the hay up, pulling it along a track set into the ridge of the barn to wherever the farmer wanted it to drop.

We didn't have such a set up, but we did have good friends with whom to chat and laugh as we pitched hay over, making a growing stack. Then it was back up to the field for another load. By the end of the afternoon we had a good stack of hay in the barn and a much greater appreciation for the baler.

We had a "new" baler that summer. Another auction find, it was a John Deere 24T throwing baler, which meant that as the bales came out of the chute, an apparatus bucked and threw the bale up and onto the rack. To use the thrower, we'd have to put sides on our hay racks. The bales deformed somewhat from being tossed up and landing on each other, so it would be harder to stack them in the barn. And it was another thing to break. After a good bit of discussion, we decided to take the thrower off and continue to use the old mechanism, which was my arms.

We couldn't possibly put up all the hay we'd need for the horses in winter with the hayloader. We took it out another time or two after that day, until the stack in the barn was well into the rafters. We needed the green baler to work.

Which it would not. Richard tinkered, bled, and swore. The thing would tie a few bales, then refuse. We'd gather up the unbaled hay as it came out of the chute, Richard would try something new, and we'd toss the hay into the mouth of the baler. We did this over and over, till the hay coming out of the chute was hard to grab, it was so chopped and short.

Someone told us about Mr. Bauer, the baler maven. I got his number and called. He arrived one morning when Richard was at work. He brought his (also elderly) friend Ed along. I showed them the baler, which we'd parked in the yard to make it easier for Mr. Bauer to work on it.

I don't remember people using the term "whisperer" to denote someone with a special knack for something back then, but Mr. Bauer was a baler whisperer. With Ed to hold his tools and provide commentary, he dove in and slowly—he was an older man, well into retirement—took it apart. I busied myself in the garden as the kids played in the sprinkler. By lunchtime the two men seemed to be feeling as if they might be done.

Wiping his hands on a rag, Mr. Bauer explained to me the necessity of always using the same brand and size and type of twine. He had a few other tips. I said I'd pass them along to Richard. "He keeps it running," I said. "I just stack the hay bales on the wagon."

The men stood motionless, as if they'd just heard something kind of crazy and had to run the tape back in their heads. Puzzled, I ran the tape back in my own head and immediately understood. It flipped through my head like a slideshow, all the times I'd seen a man and woman out baling hay—the woman ALWAYS drives the tractor. The man stacks the bales. We certainly could have done it that way, but I preferred stacking to sitting on the loud, hot tractor. I was sure I'd start daydreaming and not pay attention to the windrow, the tractor, the baler, and where Richard was standing when it was time to stop—I feared that I'd forget and cause him to fall in front of the wheel and be maimed if not killed. I'd rather stack bales.

"You stack the bales?" Mr. Bauer asked, making sure he'd heard me correctly.

"Yeah, well," I said, in as ditzy a way as I dared. "Richard says that's the easier job."

The look they shared was almost worth the hours we'd spent in the field with that shitty baler, trying to make it work. It had brought me to this moment, watching two old guys believe they hid a huge secret from me.

"Well, yeah," agreed Mr. Bauer, backing toward his pickup. "Listen, I'll let you get back to work. Good luck with that baler." He and Ed could not get into the cab fast enough, and I'm sure they had the laugh of their lives as they drove away. I imagined them telling the tale in the coffee shop for whatever years they had left, always getting a good hoot from the other retired farmers.

Oh, farming. Am I right? Sometimes you just have to make your own fun.

***

Our cat Anna thrilled the kids and me by having five adorable kittens as we watched. Two tabbies, two blacks, and one marmalade. They lived in the barn, where Anna made a nest, which the kids supplemented solicitously whether Anna wished it or not. They spent hours up there—I was stern about not touching the kittens and letting Anna be in charge because SHE IS THE MAMA. I'm sure they did sneak in some cuddles—who could resist?—but for the most part, they understood that the kittens needed Anna more than they needed to be picked up and kissed.

The kids named them, once I'd picked them up and checked under their tails to sex them. Turns out, the two identical tabbies were a male and a female, as were the two identical black ones. To know which kitten he or she had, Truman or Marian brought it to me. I'd check under the tail and state, "Mary." Or, "Tom." One morning at breakfast Truman told me he knew his cat, Billy, the marmalade one, was a boy because—and here he became very stern, wrinkling his face for emphasis—"I READ HIS NAME." I guess he thought I was reading the names of the kittens from under their tails, and it amused me to think of him studying Billy's furry little backside, deciphering the code.

Richard and I finally had all the fence pickets up, and I used spare moments to prime it. The garden flourished. The kids lived outside, usually naked unless we were baling. To celebrate Truman's potty training success, I bought a tugboat-shaped sandbox, and between that, the playhouse, the wading pool, and the sprinkler they were happily entertained. Our friends Lynn and Jay came over with their dear little girls Ellen and Jayne, and they picked out one tabby and one black kitten to take home for their own. We were left with Anna, Billy, Tom, and Mary.

One morning I was going out to catch a team in the pasture and came upon Anna's cold body, laid out flat. Something had gotten her in the night. Her kittens were old enough by then not to need her, but we were all sad to lose such a nice cat.

We also lost chickens too frequently. If we forgot, or if we were late in closing up the chicken coop, then we could almost count on one fewer bird in the morning. We set a timer to remind ourselves. One morning, though we had closed up the coop the previous evening as soon as the chickens had come back to roost, I went out to open it and found bloody carnage inside—a weasel had weaseled in and killed indiscriminately. And not even out of hunger. The dying birds had lived long enough to smear their blood on the walls, and their bodies lay scattered here and there. The remaining hens looked shocked—it must have been a horrific scene for them to watch—but soon after I'd opened the door, they were outside scratching in the dewy grass.

Years later I found a dead weasel in our garage loft, killed, it appeared, by our cat. The memory of that gory morning—the blood and the pointless killing—flashed in my mind, and I felt glad that the cat had taken revenge—though of course I do realize this was not the same weasel. Such a pretty thing, too. You wouldn't expect it to be so vicious.

***

I spent a fair bit of time that summer working on our Montessori school. We wanted to keep costs low, so we made a lot of the materials ourselves—the bead blocks, for example. We rented space from the local

Unitarian-Universalist Society. Richard volunteered as treasurer on the little school board. Meetings lasted well into the night, but he was always cheerful as he eased himself into bed in the dark.

We went out so little that small events seemed larger. The Fourth of July found us worn out, too tired to take advantage of Richard's day off either by a trip somewhere or even getting some work done. Well, we did work in the morning, getting hay unloaded and stacked in the loft. But it was a very hot, muggy day. We sank into lawn chairs in the shade of the house and watched the kids play in the sprinkler. From time to time they'd dash to the garden, pick some snap peas, and run back to us with their hands full, offering us all of them. It was exciting to feed Mom and Dad, then skitter back to the sprinkler.

That evening I read from Laura Ingalls Wilder's *Farmer Boy*, the chapter about how Almanzo and his family celebrated the Fourth of July—Independence Day. In the book, Father Wilder tells Almanzo, "It was farmers that went over the mountains, and cleared the land, and settled it, and farmed it, and hung on to their farms . . . . It was farmers who took all that country and made it America, son. Don't you ever forget that." Feeling proud, forgetting about the Dust Bowl and other things farmers have done to the land, I said to Marian, "That's right. Don't you ever forget that."

"Forget what?" she asked in some puzzlement. Wait a minute—was she even listening?

I said, with some impatience, "Don't forget that it was farmers who made America."

"Oh," she said. "I probably WILL forget that."

Once it was dark, we watched the fireworks from our bedroom windows. We could see the display over neighboring towns—small bursts in the distance over the treetops. We ran from window to window, ignoring Richard's mock complaints—he'd gone to bed and was trying, he reminded us often, with great exasperation, to get to sleep. It wasn't going to happen, not with their sharp little knees so close to his head as they knelt on his pillow and pointed joyously at every distant burst of light.

Chapter 11

We heard about an auction at a potato farm that was going out of business. It was north of us a few hours. There was a potato washer listed on the auction bill. Richard called the auctioneer to ask about it. It sounded sufficiently old, worn, and rusty to be affordable for us. So, early on the morning of the auction, we loaded our sleepy kids into the pickup and drove north.

Wisconsin is a beautiful state. Do you love open farmland? We have that. Do you love forests? Marshlands? Check and check. There are small towns crowded around lakes, towns bunched together in the midst of farm fields, towns strung along the highway with the forest at their backs. Rivers lace the countryside. It isn't unusual to see a bear as you get farther north—once I saw one sitting on his rump, legs stretched out in front, watching the cars go by. There are entirely too many deer, as evidenced by the dead ones you see on the verge.

It was a gray day, but that has never stopped an auction. The old farm was small and showed signs of having been feisty and the home of hardworking people—but on the hayracks near the house, I saw the usual progression of equipment that accompanied aging. A cane, a walker, a raised toilet seat, pairs of the expensive, stout shoes that make it possible for an elderly person to get around for a bit longer. There were stacks of old magazines—these people did not throw things out. Canning jars—I planned to bid on those, as they were pretty clean and un-chipped—and the usual assortment of dinged-up pots and pans and detritus of everyday living in a place for decades and decades.

We searched the equipment for the potato washer, not knowing what it would even look like. We finally had to ask someone, who pointed to a contraption made of once-green-painted angle iron. We studied it closely and eventually figured it out. One would dump the potatoes onto metal rollers that started at one end of the horizontal metal box. From nozzles above, water would jet down onto them. The rollers would tumble them toward the other end of the box, where they'd come out onto a series of rollers encased in sponges, which would absorb moisture and rub off any residual mud.

Upon reflection, it was pretty much what you'd expect it to look like, if you sat down to design a potato washer. Except that it was rusty, and the sponges were rubberized with age and would not absorb anything. There was no way to know if the electric motor still worked.

We learned to our dismay that someone else had bought it. Sometimes an auctioneer saves time by grouping equipment together—several lengths of chain, two spades, a rusty tire iron, for example, and maybe you want the chain. So you bid and end up with the rest of the pile. Our barn was beginning to fill with the results of this type of deal.

In this case, the washer had been grouped with other equipment. We found the guy who'd bought the stuff, and he was amenable to selling it to us for ten dollars. Richard only had a twenty; the guy conveniently (for him) did not have change, so for twenty dollars we got the washer.

It was top-heavy and sharp—something to make you ponder when you had your last tetanus shot—but we got it loaded and strapped in with the help of bystanders. More remarkably, we got it home and unloaded with, as usual, no one but Richard and me to do it. Richard set about making it useful.

***

We were looking forward to my brother Joel's wedding. Marian was to assist her older girl cousins in pouring punch at the reception, and she was quietly thrilled about it. I got out the old satin dress I'd worn in my cousin's wedding when I was six, and it fit her perfectly. I borrowed a little suit from one of the boys at the school, and Truman looked like a tiny executive in it. Or, in his words when he was trying it on and saw himself in the mirror, "I look like a prince." And he was very satisfied about that, too.

On an evening a few days before the wedding, the phone rang and I answered it. My sister-in-law, Gayle, was calling.

"Maureen? This is Gayle."

"Oh, hi! How are you?"

Pause. "I'm calling to tell you that your dad has had an accident." My breath stopped. "He's ok, he's not dead. But your mom wanted me to call and tell you that he's going to be in the hospital."

Well, that didn't sound so bad. I imagined a fender bender with the car, Dad hurting his knee or something.

"What happened?"

Gayle explained that Dad had been driving the tractor back to the farm with a chopper box. He'd made a left turn to go in at the farm driveway, and a loaded semi-truck with trailer had hit the tractor broadside, splitting the tractor in half. It was the farm's first tractor with a cab—otherwise he'd have been unshielded from the impact.

I tried to imagine what kind of injury this would NOT cause—was there anything left of him?

"Scott (my youngest brother) saw it happen, and he ran to him right away. Then he ran for me. When I got to him, he wasn't breathing. But we restored the airway and got the ambulance there. And someone went for your mom."

Not breathing?

My sister and my mom took that first day and night with Dad at the hospital. He was confused, wanted to go home, couldn't stand being in the hospital, could not be reasoned with.

I took the second day and night. My brothers did their parts, all five of them, plus my foster brother, Gerald, who took a long shift on his own one day. Joel's wedding was coming up. Dad talked about it as if nothing had happened, as if he'd be attending.

"You'll get your hair done," he told my mom, reciting the steps they'd obviously worked out before the accident. "And then we'll pick up the little guy (my nephew Roy, who would be in the wedding) and drive down to Gladys's (my mom's sister). And she'll come with us the rest of the way."

He had it all down pat, as my mom confirmed. But he couldn't place himself in the moment, couldn't understand what had happened to him. He'd try to get up, we'd try to reason with him, he'd lie back and listen to us talk for a while, then do his old polite-farmer-getting-out-of-a-boring-situation schtick: "Well," he'd sigh, as if reluctant to leave. "I suppose." And he'd struggle to get out of bed, and we'd start all over again.

He was sleeping when my mom pointed out the scratches on his arms to me. "They thought those were from the accident," she said. "But they weren't. He was picking blackberries beside the railroad track with Roy and Carter." Carter was another of my small nephews. I could see them, my dad and his eager, hopping grandsons, Dad picking and the boys eating—Dad's idea of a perfect day.

"This isn't the end," I said to my sister on the phone. "But I think it's the beginning of the end."

Richard and I had planned to leave for Italy—a work trip for him—from the wedding reception. The kids were to be stashed with friends. I was looking forward to the trip, but of course I couldn't go. I didn't think about it much, in fact. When Richard called from Como to tell me he was there safely, he said that it was beautiful and that I would probably really like it there. I just let those words go through my head. I spent the wedding reception at the hospital with Dad, who had started breathing like a freight train. I thought he'd die in front of me.

He didn't, though. He was in the hospital for a few weeks. Truman and I had a routine on Monday and Friday—drop Marian at school, then drive the one-hour trip to North Memorial and spend the morning with Dad. Truman patiently played with his trucks and tractors on the floor of Dad's room. Then we drove home to pick up Marian before trying to pick up the pieces on the farm. On Wednesdays I left Truman with my friend Lynn—Truman and her daughter Jayne were friends, and they went to a preschool program together. However sad I was from watching my strong, independent dad struggle to answer the inane questions posed by the occupational therapist, I was always glad to spend a few minutes with Lynn, hearing about the funny things Jayne and Truman had done together that day and feeling the hospital slip off my shoulders.

When Richard came home from Italy, we began digging potatoes in earnest. Now that we had the washer up and running, plus the new storage room, we were feeling cocky about our farming capabilities. We'd put up hundreds of bales of hay, enough to feed our herd of more than twenty horses through the winter, doing all the mowing and raking with horses, a few of which we had raised and trained ourselves. My garden was feeding us, as long as we mostly wanted to eat tomatoes and beans with our potatoes. For days on end my kitchen was a canning factory as I put up tomato sauce to use on spaghetti all winter, and the kids were pressed into service, snapping beans and husking corn for me to blanch and freeze.

Richard had found replacement sponges for the washer, had cleared the spigots so water would jet onto the potatoes rolling past, and had built a sloped table for the cleaned spuds emerging from the washer to be spread out and examined for defects or green spots as they air dried and then were packed into boxes to be carried to the new storeroom and stacked on pallets.

We dug several rows of potatoes with the potato digger and picked them up into our stackable crates, which we hefted onto a hayrack. There was always a good bit of calculation that went into how many rows we would dig—we had to take into account how late we'd gotten started and how long it would take to get the potatoes picked up before dark, or before we had to get to a meeting, usually about the Montessori school. If we knew it would not freeze, and there hadn't been much depredation from animals on potatoes left aboveground, we dug more rows than we could pick up, and next day when Richard was at work, the kids and I went up and worked. Then, when Richard got home, the lucky man got to put on his farm clothes and stand at the end of the washer as I tipped half a crate of potatoes at a time into it. Usually I hosed them first, while they were in the crates, so before very long I was slimed with mud. If the day was cold, I was cold.

The washer did a fair job—much better than any of my previous efforts.

Richard, once the potatoes started coming through, had a busy time of it, sorting and packing and labeling the boxes with the grade and variety of potato inside. At first we tried switching off, but the person putting the unwashed potatoes into the washer got so muddy that inevitably it got onto the clean potatoes, and they had to be wiped or put through the washer again.

Our mare Belle was due to foal soon. She was a squat horse, and late pregnancy had only added to the impression she gave of being a dark-red chest freezer on legs. I watched her udder closely, as I worried about her ability to get that foal out of herself. It was her first, and she was such a meek horse, always so careful not to get in the way of the rest of the herd.

"Mama, can I sleep in the barn with you and see Belle have her baby?" Marian asked me one day as she watched me lean over and eyeball Belle's underbelly.

I couldn't see a reason why not. "Okay," I said. Truman chimed in that he wanted to sleep out there, too. We stacked bales on an empty hayrack near the foaling stall and put out our sleeping bags and pillows. That night we went out and found Billy, Tom, and Mary snuggled into our beds, which delighted the kids, of course. They went to sleep easily, and so did I.

The sound of Belle's water breaking woke me up. I turned on a light over Richard's shop bench so I could see better, but Belle would still be in relative darkness. She was definitely in labor—sweating, lying down, getting up, turning in tight circles. I woke Marian to see if she wanted to watch. She did, and she blinked industriously to wake herself enough to take things in. I shook Truman a few times as well, but he could not be awakened.

I always worried when a horse was laboring. The memories of Lucky Romper and Superman were strong. This was taking too long.

"I'm going to get Daddy," I whispered to Marian, who nodded as I slipped from the hayrack and into the darkness outside the barn. She had not moved when I got back a short while later.

"Still nothing?" Richard asked when he arrived. Nothing. "Maybe you should check in there," he suggested.

I'd been thinking that myself. I put on one of the long, disposable plastic gloves we kept in the foaling kit for this purpose and, while Belle was lying down, panting with effort, I slipped my hand into her vagina.

I felt the hooves, as I'd expected. But it was odd—they were upside down. This foal was bring born not as if he were diving into the world, which is the normal presentation, but as if he were doing the back float.

"Call Peter," I told Richard. "I think it's stuck."

We had a portable phone that sometimes worked even up at the barn, and Richard had foresightedly brought it out with him. It worked, amazingly, and he reached our sleepy vet. Questions went back and forth, frustrating me—I wanted that foal out, right now. "Just tell him to come," I said to Richard.

Peter came and also felt inside Belle, coming to the same conclusion. He, however, knew what to do, and he explained as he did it. He checked to be sure the foal's nose was positioned between the fetlocks and would not be bent back as it was born. He got foaling chains out of his kit and fitted them, working blind inside Belle, onto the foal's forelegs above the fetlocks. He made sure the hooves were offset. Then, next time Belle labored, groaning with an effort so great that it rocked her onto her side, legs stuck out straight and stiff, he and I pulled downward, toward her hocks. Nothing. Belle got up. She lay down. Again, labor clenched her body. We pulled, and this time a hoof appeared, and then another, and a wee nose with a dark purple tongue sticking out of the mouth. The color of that tongue convinced me that the foal was dead.

I braced and pulled again the next time she had a contraction—this time she was standing, and Peter reminded me to pull toward her hocks. The foal slid out, landing on us with a slippery plop.

He panted, scaring me that he was another Superman. But he eventually stopped that and breathed evenly, looking around. Belle touched him with her nose, whickering the unmistakable low notes of a mare to her newborn foal. Eventually he got up and started wobbling around on his ridiculous legs, then began the search for her udder. We laughed when he solved the problem by standing behind her and sucking from between her back legs. None of us had seen that before.

Next morning I called my friend Inga and told her about the birth. "You should name him Tug," she suggested. "How did Marian take the whole thing? Did you just life-experience yourself out of ever getting grandkids from her?"

We laughed about it, and Richard and I agreed that Tug was a great name for the little colt. As for Marian's reaction to the difficult birth, luckily she did not seem to connect it to something that might ever happen to her. Though she was awfully quiet about it.

Truman slept through the whole thing, covered with cats.

***

The Montessori school was a reality. We had six students and a teacher. We'd made a lot of the materials and purchased what we could not make. The evening before the first day of school, our six families and the teacher met for a picnic. Later, we gathered in a circle and each parent said something about their thoughts regarding the year ahead. Everyone was hopeful and optimistic. Tim, one of the dads, made us laugh when it was his turn. He simply turned to Ellen, the teacher, and said, with a glance at the kids, "Better you than me."

Our rented space was used for Unitarian-Universalist service on Sunday mornings. So Sunday evenings we had a rotating calendar of parents who had to go in and move the pews—old, wooden collapsible movie-theater seats (the UU folks didn't have much more money than we did)—and haul out the shelves, arrange them with the learning materials that had been boxed up to be stored, and, in short, make the UU space into a Montessori space. Plus clean it. One of the moms, Nanette, had devised a checklist that I found illuminating. I'd never considered washing a light-switch plate. It made sense, as did the washing of the doorknobs. Sometimes I felt as if I'd missed some important information along the way.

On Friday evenings we did it all in reverse, making the space ready for UU services on Sunday morning.

Marian settled into school very happily. She was the only girl, but that didn't bother her. The other students were well known to her—we'd been getting together all summer to work on the project. The kids had done some of the work themselves. It made the school seem special to us all.

I hoped she would learn to spell. She was reading well enough for her age, and she kept a little journal of the books she read. She left notes all over the house for Richard and me: I lof eeuuoo. She was sounding out words as if she heard them in slow motion.

One afternoon Richard and I came down from the field and found, in the house, not our children—which is what we'd been expecting to find—but a crinkly note on the table: *Dad Wer in the wods. We fond fiznt eggs.*

We walked down to the woods and "fond" them easily enough, and they brought us to the leaf-camouflaged nest of pheasant (or whatever bird it was that laid them) eggs. They hadn't touched them, of course. They knew better.

It began to seem as if one or the other of us was away from home that fall. Richard had to go back to Europe for a week in November. My dad was back in the hospital, so I was away and visiting him when I could. Friends stepped in to help with picking Marian up at school or taking Truman for the day when it seemed as if he just could not take another long drive to Minneapolis and a stint of several hours at the hospital. When he did come with me, he brought his toys and played with my mother or whichever of my siblings might happen to be there.

My dad's brain was bleeding again, and after the surgery to correct that, they found cancer in his bladder. He had a surgery to remove it. It seemed as if once he'd been brought down, ill health was going to just keep on kicking him. Marian asked to see him, and I brought her in on a Saturday morning. She clambered into bed beside him, leaned her head on his chest, and read to him from the book she'd brought along for that purpose.

Potato harvest continued, with marathon sessions of digging, picking them up, and then washing and sorting and boxing. Sometimes I could drop orders at co-ops in the Twin Cities on my way to visit Dad at the hospital. I worked at winterizing the garden in little spurts of stolen effort. The kids helped, sometimes just by being their silly selves. I came in one day and found Richard sitting at the table, reading the paper as both kids—each one standing on a chair beside him, each one with a comb and a glass of water to keep dipping it into—carefully combed his hair. All three had trancelike looks about them; "Looks like you have a crack in your head," Truman noted soberly, rubbing at the gleam of light reflecting off his dad's balding scalp.

Another day the phone rang, and as I picked it up to answer, Truman said, "Know what, Mom? My butt is sad." Luckily it was Richard on the other end of the phone as I sputtered "hello" and tried not to burst out laughing.

Because we had enough hay to get through the winter and a barn with a stall in it for a foaling mare or sick horse, we were happy when snow began to fall that November. We had the potatoes safely stowed in the storage room, and Richard had even constructed a V-shaped wooden plow with which he planned to keep the road open. I went for a run late in the afternoon just as the snow was starting to fall. Richard was home with the kids. When I got back, I started shoveling the apron in front of our garage.

"Where's Mom?" Truman asked. Richard told him I was shoveling snow, which galvanized our little man, and he was outside with his little shovel within minutes. Apparently he had a thing about snow falling on the grass of our yard, so he set to work to clear it. I finished my work and went in, and we watched him push his shovel, clearing a strip of grass, from one end of the yard to the other. By which time the snow had covered the grass again, but this didn't seem to bother him. He just kept working until I called him to come in for dinner.

As I helped him get out of his snowsuit and boots, I said, "There's still a lot of snow on the grass." He sighed in weary agreement. "I'll get it tomorrow," he said.

The snow made Christmas seem imminent, though it was still a month away. We drove a bobsled up to the woods to pick out a tree—a skinny cedar—and dragged it back to the living room. Friends came for Christmas Day, and we hitched Katrina and Gidget to pull the bobsled around the farm, dragging the kids behind us on a sled tied to the runners with long ropes of twine. January was again very cold, but, happily, no mares were due to foal, and we could stay in our own beds at night.

Though one night, when it was nearly 40 below zero, the moon was full in a clear sky. Some of the stars—probably planets—were bright enough that I could see them from my pillow, without my glasses. It was so cold, and I worried that the horses were suffering.

I tried to calm myself—it was a dry cold, they were furry, they were getting plenty to eat, there was no wind. But nothing helped, and finally I got up and dressed and went out to find them. At the gate I started calling for Katrina, the horse I felt was most likely to respond. Nothing. No sound. Just the staticky brush of my hat against my ears. I walked out into the woods and jumped at the sight of what, for a split second, I thought was a snake. As if a snake would come out and slither over the snow! It was a stick, dark against the white forest floor.

No horses. Then I realized that I had two thick hats on and was, essentially, deaf. The horses could be coming, for all could tell. I ripped the hats off my head and sure enough, I could hear the pound of their hooves as they galloped up out of the trees. I dodged over to stand near a sturdy one and waited—the horses emerged, slowing to a trot, steaming from their nostrils and looking perfectly fine. In the moonlight I could see their curious looks and I felt silly for having disturbed them. They followed me up to the barn and I threw a bale of hay to them, taking care to separate it into flakes so they wouldn't have to fight over it.

Back in bed, I was careful not to touch Richard with any chilled part of me. I didn't want him to know how silly I'd been.

Chapter 12

I got off the plane at the Seattle-Tacoma airport and scanned the crowd for my friend Peggy. Tall and red-haired, she was easy to spot. "Mo!" she hollered, waving me her way. Our hug was long and tight. It had been too long since we'd been together.

"It takes about an hour to get to the farm from here," she explained as we got into her car. "Do you want to go somewhere for a bite to eat, or just head out to Machias?"

"I want to see your place. You'll have something there to eat, I'm sure." I was eager to see the homestead she and Mike had found after a year or more of looking. She already had a couple of sheep, Mike was building raised beds for gardening, and her letters to me were full of news about lambs being born and pear butter being made. My big-city friend!

It was getting dark when we pulled in, but I could see the outline of her cute story-and-a-half bungalow house. The pines blocked out the rising moon.

"There are some chickens roosting in this holly tree," she told me as we hauled my stuff to her back door.

As she had so often in the past, Peggy made dinner for me as I sat and enjoyed just being around her. She had a way of making whatever space she inhabited seem fun and funky. This old-timey kitchen was no exception.

"I looked up how far it is to Portland," she said. "Where exactly is this horse you want to see?"

The reason I was visiting her was first of all to visit her, of course, but there was some mission creep—Richard and I needed a new stallion. The fillies Ezra had sired on our farm were now coming up on three years old and were ready to be bred. We needed a stallion unrelated to them, and we'd located one in Oregon. Just a state-and-a-half away from Peggy's house—no big deal.

According to Peggy, it wasn't. We took off the next afternoon after I had a chance to see her little farm in the daylight, to meet Blossom and Florence—her two ewes—and to admire her neatly planted raised beds. We got on Highway 5 and headed south.

Peggy and I met at our first jobs after college, the local weekly newspaper in my hometown. I was a bit late to work my first morning—as I did not have a car, I rode my bike from Mom and Dad's farm outside town. I'd broken my foot a couple of weeks earlier and it was in a walking cast, which made it difficult to stop and start on the bike. For that reason—and because, really, is it necessary to stop at the sign when you can clearly see no one is coming from either direction and you are on a bike in Milaca, Minnesota?—I just whizzed across Main Street and, what do you know, I hadn't noticed the squad car in the alley.

No ticket, just a warning, but I was irritated when I stumped through the door of the Times office, dangling my helmet from one hand and shrugging out of my backpack. There was a woman about my age, it seemed, working at a tilted surface in the space beyond the office. The look she gave me was kind. Later, I learned, she'd thought I was mentally challenged.

The thing about coming back to your hometown after failing to accomplish the thing you'd always said you planned to do—go to medical school—is that it is embarrassing. And lonely. Your friends are off doing the things they said they were going to do. They did not fail. You did. The only people left in town who are your age are married with children. Nice people, but busy in ways that are just not that interesting to you. You do your stupid little job at the newspaper and watch the woman who is around your age do hers—which is a real job, the job she went to school to learn—and you find ways to get to know her.

"Oh, yeah, I swim," Peggy said one day at lunch. And that was it. I loved swimming. The pool at the high school was open for adult lap swimming in the morning, and what made more sense than that the two of us would go there together and then, from there, come to work?

And so began the legend of Peggy and Maureen. We ended up living in an old house together, cleaning out the dead mice from the silverware drawer and using a broom to dislodge the cobwebs from ceiling corners in all the rooms. That winter there was a snowstorm every weekend, and we stocked up on M&Ms on the way home from work (Peggy had a car, I still did not) on Friday nights as if they were staples, as if they were one of the four food groups. In a way, I guess they were. Peggy's car was a baby-blue 1975 Satellite Sebrig, jacked up in the back and always in need of a 1) a jump from Harold, our elderly neighbor across the gravel road, and 2) being shoveled out, right down to the dirt, it had such bald tires.

As we waited for Harold to finish his coffee on cold mornings when the car wouldn't start, we'd go back into the house and Peggy would pull out the other three food groups—flour tortillas, cinnamon sugar, and butter. The skillet was always on the stove, and it was no trouble at all to get the tortilla hot, butter it, shake on the sugary delight, roll it up, and tip our heads back to enjoy the pleasures of cinnamon sugar, flour, and fat.

Wishing we had time for three or four more of those, we put on our snow clothes and went back out to the car and started shoveling. Harold would show up with his pickup and jumper cables and get us going yet again, and off we'd go to the office, where Peggy was the reporter/photographer and I laid out ads.

Even the pleasures of sharing a freezing old farmhouse with Peggy, her extravagantly furry cat, and her tiny, blurry, black-and-white TV set were not enough to keep me content, and I diligently used as much time as possible at work to apply to jobs elsewhere. Finally, in March of that year, I landed the one at the magazine in Illinois and left Peggy to face the spring Flooding of the Basement by herself. We never lived in the same state, let alone house, again after that. It's possible the U.S. Postal Service noticed the bump in stamp sales once Peggy and I were apart—we were both letter writers and had found in one another a person who enjoyed hearing our take on the day's events. It does not get much better than that, I believe.

Now, years later, as usual it was Peggy at the wheel, driving us toward Portland, where I wanted to visit Klickitat Street in honor of Beverly Cleary, the children's author I so admired. That evening, as Peggy nosed the car onto a frontage road where we were to stay at a motel, a gaggle of teenage girls appeared in their strappy sandals, long white legs, cut-off shorts, tight tank tops, and artfully disarranged plaid flannel shirts. I won't even go into the amount of make-up they were wearing. It was almost painful to see how hard they were working to look casual, and we both grimaced in sympathy. "Trouble on the hoof," Peggy commented.

On that trip I visited a young stallion prospect and took pictures to show Richard when I got back. We also stopped at a farm to see a mare we were interested in, and I made arrangements to trade two young geldings for her. Diana was a dark chestnut mare, blind on the left side, with a stocky build and bloodlines that would be new in the Midwest. Suffolk horses are rare enough that it would be easy to lose genetic diversity. Luckily, we'd found, owners of Suffolk horses were careful about breeding out whenever they could. There was still a lot of variation in the breed.

It was wonderful to be with Peggy, but I missed the kids and Richard and was happy to get home. I wrote to thank her for the great time and for driving me hither and yon, and to tell her how much I liked her new place. "The only thing I didn't care for was that it is on a busy road," I wrote. "But there isn't much road noise." Then, after she'd written back at some length about the birth of Florence and Blossom's lambs, I wrote in response, "When I compare your story of the birth of the lambs to the story of a normal horse birth, they are very similar. There is one difference, though, and that is the sense of history one has in dealing with a foal. A foal is like a child, in a way—it has a life of possibilities stretching before it. When I rock back on my heels and look the wretched thing over, I see the future. If I raised sheep or cattle, would I look at the newborn and think about how it was going to taste? On the other hand, we can't all raise horses."

I went on from there to other topics, put the letter in an envelope and mailed it to Peggy, and carried on in my arrogant ways, never suspecting that I had said anything that might be taken amiss.

A week or so later I eagerly ripped open a letter from Peggy—always welcomed, always a good read. "I've been thinking about writing this letter all day, wondering if I should write it, deciding I should, deciding I shouldn't. Obviously I have decided to do it."

What? I continued to read.

"I savor your letters. But a couple of comments in your recent letters have struck me as smug and spiteful. I wish you'd think a little more before you make a point of mentioning that I live on a "busy road" or conclude your discourse on the limited life spans of our sheep with "we can't all raise horses."

Well. She went on from there in the most delicate and tactful of ways, my great friend, correcting and praising me in the same sentence, pointing out the ways in which I was being an ass and how she admired me all the same.

I couldn't believe I had written those things, but I went back to the floppy disc on which I had saved the letters and read with some sinking of my stomach that indeed I had. Indeed I had.

I did not choke on the crow I ate that day, but I might have. Should have. I quickly wrote to apologize and make amends, if I could. My great-hearted friend forgave me, and I had a good and much-needed lesson. I was cocky about our endeavor—it was so cool and interesting to me, I thought everyone shared my sense of wonder about it. Who knows who else I insulted or irritated? I shudder to think about it.

"What if I wrote to you something on the order of: 'Not everyone wants to live in the middle of nowhere on a wind-battered rise, surrounded by mud, but you are doing the best you can with what you have, I guess'" Peggy had written. Followed by, "Which I wouldn't have written because I don't think it."

She might well have thought it, though. It was an accurate description of our farm, especially that March. The snow melted very quickly, the frozen ground did not, and our horses churned it into a sucking mess up near the barn.

I was checking sap buckets one afternoon when I heard what sounded like a huge plane overhead. That is not what it was—rather, it was the horses galloping back from the far pasture. The beat of their hooves was a shade slower than usual because they had to pull each foot out of the mud every time they moved, and there was the accompanying sound of sucking. Each strike kicked up four globs of mud, and they moved in a hail of it.

I dreaded feeding them, and I had to do it twice each day. They were hungry and pressed against me as I came from the barn with their hay. In winter, I loaded a sled with a couple of bales at a time, and I had the top bale ready to spread out right away, thus distracting them so I could spread out the rest unmolested. In the deep mud, I could only take one bale out at a time, and that first one always felt as if it might be my last. Unsteady because of the sucking mud, I couldn't move quickly without losing a boot to it. The horses jostled to get a snatch of hay. They were respectful of me, but not of each other, and as the horses with lower status dodged the nips and threats of the more powerful horses, I was vulnerable.

I couldn't just throw down a bale and run—it would so quickly be pounded into the mud. I had to find a clean, relatively dry place to put the bale down, cut the twine, and spread it out. Then go back for another and another and another and another. Gluck gluck gluck, all the way back to the barn and all the way out. Once the first bale was down, I was mostly in the clear. But I was already dreading the next feeding.

We tried to imagine a better system—a long feed bunker, feeding them from a wagon, a catapult that would pitch the bales far into the pasture, beyond the mud. By the time we were in a position to make improvements on the eat-off-the-ground system, the mud was gone and we were deep into spring work and put it off till next year. Every year.

The weanling stallion and Diana, the pregnant mare, arrived from Oregon. In walking Diana around the pasture to familiarize her with the fences, Richard stepped into a deep puddle and got a good dose of manure water—I called it farm tea—in his boots. I was laughing at him when she tore the lead rope from his hands and leapt away, churning up mud divots that landed all over me. I wrote to Peggy about it. "See? Not everyone can raise horses. Most people are smarter than that."

We made ten gallons of maple syrup from our box elders that spring. Luckily, we'd bought a larger pan for the boiling—six feet by two feet—and we supported it on a box made of cement blocks. I'm sure someone more discerning might taste a difference between the syrup we made and syrup made from actual sugar maples, but we did not. We were grateful to have the syrup, and after all the hauling and boiling and keeping the fire stoked, I was glad to be done. Bringing the last of the buckets down the hill on a sled drawn by Gidget and Stella, I noticed that the pickets of our still-uncompleted fence around the yard would be useful as places to tip the washed buckets as they dried.

We had a new and compelling reason to finish the fence. Back in January, Richard had been on the phone with Bill, a friendly veterinarian from southern Wisconsin and one of the directors of the American Suffolk Horse Association. Bill said that the association needed a place to hold its annual gathering and meeting the next late August or early September. It should be a place within an hour of an airport, with lodging available in the nearby town, and on a farm with a good number of Suffolk horses.

Richard finally grasped that Bill was suggesting our place as being perfect for that year's gathering, and he was surprised enough to agree. Suddenly we were going to have a pretty big event on our piddly farm, and I wanted to make a good impression.

As big events go, this gathering was nothing much. But we wanted our horses to look good, our yard to look nice, and our fields to be relatively free of weeds. Not so easy at that time of year. I once heard Garrison Keillor, the radio personality, say that August is the month when summer starts to seem like an experiment that didn't work out. By August, the weeds that had not been caught by earlier cultivations were rank and strong. They could only be conquered by hand. The potatoes were starting to die back, which allowed sunlight to reach and encourage the weeds beneath the vines. The wheat would have been harvested by then. I'd be deep into canning garden produce. Our best bet for getting ready for the event would be to start early in the springtime.

***

Katrina bagged up and I put her in the nursery paddock so I could watch her during the day. I slept out with her at night—for eight nights. She had an enormous colt after an exhausting labor—he was jazzed and ready to go, it seemed, right away. She lay on her side, gasping for ten or fifteen minutes after the birth. Then she raised her head to look back at her new son, and I felt her fall in love.

He was a husky fellow, trotting circles around her within an hour of standing up. The kids and I went out to visit them the next morning and noticed he had a little cowlick on his side in an unusual place, with a few white hairs in it. "It's like his mama kissed him," observed Truman. So we named the foal Kiss-Me-over-the-Garden-Gate, for a flower I was hoping to grow that year.

The calendar that April was filled with breeding notes. Jemima in heat. Stella out. Belle in. Bred Rosalie-Carl. Gidget-Ezra. And so on. We had ten mares to get pregnant. I began making noises about too many horses—we were not selling them very fast, and with so many young horses, we could not keep up with getting them halter broken or used to having their feet picked up, or even, in many cases, easily caught. It bothered me a lot.

Our living room was crowded with seedlings that tipped toward the sun. Every day the kids and I watered and turned the flats of tomato, marigold, squash, onion, leek, and assorted other seedlings. I ran a fan on them for an hour or so every day, preparing them for the reality of growing on our windswept farm. As the weather warmed and the plants grew larger and needed to be re-potted and took up even more space, I devised cold frames outside, using straw bales and old windows I found at our town's recycling center.

We were all itching to get out and get started with the growing season. Richard hated leaving, but he had another trip to Europe lined up. Two weeks during this busy time of year—it was painful. "At least I got the wheat planted," he sighed as we drove him to the airport. He hauled himself out of the car. "We'll be ok," I said, not feeling it. He slid the side door open and gave each of the kids a kiss and a hug good-bye. As I waited for my turn to pull out into the line of traffic heading for the exit, I watched my dear husband's back disappear into the airport and thought these next couple of weeks would be very, very long.

Richard had built window boxes as Mother's Day gifts for me that year, and I positioned them in each of the dormer windows. The kids and I hunted for morels in our woods, finding only a few but delighted most of all with the wild flowers we had not known were there. Showy orchis was a new one on me—it was hard to believe such an exotic and delicate flower could grow on our farm. A leaf-matted slope down to a ravine seemed entirely too prosaic a place for such a beautiful thing. But that's springtime for you. Violets peeking around a pile of horse manure, bloodroot flowers gleaming their waxy whiteness beside a brown and rotting clot of snow. We learned that Truman was best at finding morels, whether because he was shortest and therefore nearest the ground or because he had the sharpest eyes—that question would not be answered for years. In the meantime, he and Marian loved hunting morels and didn't care to eat them. It was a perfect combination for me.

The long weeks of Richard's absence passed, and he had the pleasure of the kids going bonkers when they saw him. At home he gave out the usual gifts—the accumulated shower gels, shampoos, caps, fancy little soaps, and, sometimes, the chocolates that had been left on his pillow in the fancier hotels. Marian and Truman enjoyed lining them up in the bathroom cupboard and choosing which soap and shampoo they'd use next. We certainly never needed to buy any of those items, and we had plenty to give away.

"Thank you, Daddy!" they exclaimed, sorting through their loot, and Daddy and I exchanged our amused glances over their heads. "Of course," he said, magnanimously. "I'm glad you like it."

Richard was sick when he came home, and for the first time since I'd known him, he took a couple of days off work. I told my mother about it on the phone.

"I HOPE he doesn't have Ebola," she said. Right, Mom. She kept up with the news almost too closely. When I mentioned a gay friend of mine and said that we'd planned to get together but he came down with a sore throat, she said, darkly, "I HOPE it's just a sore throat." She paid little attention to her own health, which was terrible due to the rheumatoid arthritis she'd been afflicted with in her mid-fifties. Her own pain and limited mobility were far less on her mind than these unlikely scenarios that could threaten the health of her children, grandchildren, and anyone else in her orbit.

"How is that fence coming?" she asked. She was concerned that our farm might not look its best for the Suffolk Association gathering in late August. "Dad and I can come down and help one day next week. We can paint." Dad was well enough to do light work and Mom saw how he needed to feel useful.

Turned out, Richard did NOT have Ebola, lucky for us. He recovered and we got the potatoes planted and began making hay. My parents came down and helped us prime the fence, once Richard had the final pickets screwed on. "We'll be back to help paint it," they promised, kissing the kids good-bye.

It was an astonishingly hot summer. We harnessed the horses to cultivate the potatoes well before the sun came up. I'd be out in the field watching the light grow stronger to the east. I could be driving the team down a row, paying attention to only the plants below me and the horses in front of me, and I'd know the sun had just breasted the horizon, its heat was so immediate. The horses quickly darkened with sweat, and our rest breaks were ticked off in droplets of it falling from their bellies.

We had to wait till the dew was off the grass before we could mow hay. I used Katrina and Lucinda, who walked fastest and seemed to tire most slowly. They breathed like locomotives, even their beautiful faces darkening with sweat. I rested them often. Katrina worried about her foal, Kiss-Me, who was in the pasture with the other foals and mares. It was funny to me how the mares seemed to understand that a foal left in the pasture while its mother was working had no choice but to wander around and try to play with their babies. Generally, until their foals were old enough to go off and play with the group, they didn't like another baby coming around to check on theirs. The most forward colt I'd ever seen was Tug, who nursed from other mares at will, just pestering them until they finally gave up and let him suck. He continued to nurse from his mother, Belle, from directly behind her, and it made us laugh to see him stop and peer at us, his mother's tail draped over his head like a stringy hat.

We sold Martin VanHuisen. He was a friendly, well-built horse, and he went off to be a stallion for someone in northern Wisconsin. A man in New York bought Gidget, and an acquaintance of ours in Minnesota bought Spike, also intending to use him as a stallion. We were happy with how the weanling we'd bought in Oregon was looking. His registered name was Centaur's Diamond, which didn't really roll easily off the tongue.

"Diamond" was the obvious choice, but that was the name the kids had given the stray cat living in the barn. We tried several names starting with D; because it sort of fit him, we lingered on Doofus, but discarded that as being bad for his self-esteem. Finally we tried on "Damon," and that was it—though if either of us referred to Doofus, we both knew who was being discussed. He was in-your-face friendly, with thick legs and a tendency to take up his own space as well as that of whoever was standing next to him, whether horse or human. It was cute when he was smaller, but of course we knew he would get a lot bigger. Whenever we worked with him, we stressed PERSONAL SPACE boundaries.

***

"I have an idea," Richard said, pushing the shower curtain back after he'd cleaned up from a hard afternoon outside. I was putting on a clean shirt, preparatory to making dinner—no one liked horse hair in the food. For once, we were not working through dinner and planned to spend the evening taking the kids to a movie.

"What if," he said, still deep in thought, flipping the towel up over his head and down behind him, holding it at the ends and drying his back in a slow, methodical seesaw, "we set up a pedal-powered grain mill on a trailer, and ground flour with a bicycle for people after they bought our wheat at the farmer's market?"

I mention this scene for two reasons—one, the complete ridiculousness of my wet, naked husband, beard still dripping water onto his pale chest, pondering and drying, drying and pondering, thinking—as usual—of more work we could do. And, two, to note that we were still growing wheat and had high hopes of adding value to it. And, even all these years later, I still think that is a good idea he had, but we have so far not followed through on it.

We planted sorghum and canola—just an acre of the sorghum and about three of the canola. Neither was a common crop in the region, but we knew we couldn't compete with conventional farmers on the usual playing field of soybeans and corn. We liked the idea of making our own sorghum syrup and canola oil.

The sorghum came up looking like corn. The canola came up in a promising fashion, too. It began to seem as if we were on to something.

We baled more than one thousand bales that hot June. Richard kept watch on me from the tractor seat, waggling the canteen to see if I needed to stop for a drink, making sure I didn't get too far behind, with bales piling up on the wagon. When the shear pin on the baler broke, as it often did, I slid off the wagon and found shade beside it and rested as Richard made the repair. One afternoon, one of the big tractor tires went flat. It was going to rain, and we had a small field of hay down. We rolled out the hay loader, hitched three horses to the forecart, and enlisted the kids.

They rode with me on the forecart. Richard placed the hay as it cascaded over the loader and onto the wagon. When I stopped the team to rest them, he hopped off to hold the horses while the kids and I climbed onto the load to tromp it down. The air grew heavier, the sky grew darker, and we pressed that hay tightly, hoping to get the whole field of it onto one load.

There might be a better feeling than the one you have when you close the barn door, protecting your newly loaded or baled hay just as the rain begins to tap and then drum on the roof. I can't think of it just now.

That day, as Richard put the chains on the door to keep it from flapping out in the winds that were sure to accompany the newborn storm, the four of us were as wet with sweat as the horses, covered with hay bits, sagging with effort. And happy. The kids—tired as they were—puffed up with having been truly helpful to us. We never were a high-five kind of family, and our arms were too tired that day, anyway. But it was a high-five kind of feeling.

***

Baling Hay

We used to make a lot of hay. He drove
the tractor and I stacked
bales in an intricate pattern that held them
even as the wagon rocked on our uneven fields.

I never liked pulling on jeans in hot weather,
or work boots, or the prospect
of all that work ahead.
Had to be done, though, and we hitched

tractor to baler to wagon and I hopped up, hay hook
in hand, knees bending to absorb the roll
as we bumped out to where what had been grass
was now hay in long billows
striping the cropped field.

Most husbands stack the bales, wife drives
the tractor but we did it the opposite way
and for all my reluctance to start, the first
bump of the bale coming out of the chute
seemed like a good omen of harvest and lifted
my heart, or maybe it was the clear pure task
ahead--making my careful, square stack—or
knowing he'd slow down, even stop, get down
and help if I got behind in my work,
waggle the canteen at me—thirsty?—
from the tractor seat,
maybe remembering
one time
baling our way past four newly fledged
bluebirds sitting on the fence wire, the way
their heads turned soberly, interestedly,
as we clanked past
and I looked from them to him, on the tractor,
who at the same time shifted
his gaze from them to me, a brief raise
of his eyebrows, our small baling amusement,
what passes for entertainment,
bird and human, in a hay field,

maybe those hours when an old, finicky,
clunking blue Ford baler separated us but
the task united us under the sky, over
our farm fields, and after we'd got the hay
safe in the barn, showered off most of the dust,
fed the kids put them to bed turned out
the lights and the breeze blew over us
we fell asleep as if we'd been baled ourselves,
heavy and solid, dreamily swaying on our sheeted wagon,
our careful stack we have held now for years.

***

My family had a celebration for my parents' fiftieth wedding anniversary that weekend. We drove up and joined with my siblings, their spouses, nieces and nephews, neighbors and friends in applauding Mom and Dad's long marriage and contributions to their community. My dad was thin, and the surgery on his brain had left a dent in his forehead, but it was good to see him quietly enjoying himself. Nothing made him happier than being with his grandchildren. My mother made him wear a suit, which he hated, but he soon also wore a small child or two, holding them on his lap when seated, or hugging them close when standing. That was always his preferred attire.

They'd asked for no gifts, but of course that was not going to happen. My sister set out a basket in the church reception hall for cards from well-wishers, and the next day Mom and Dad sat down to open them.

"Look at this, Roy," Mom said, handing Dad a card. "From Sandra. Isn't that pretty!?" And Dad nodded.

Often there was money enclosed—Mom couldn't get over it. "Five dollars!" she exclaimed. "From Mrs. Nelson! Oh, that's too much, she shouldn't have done that!" And, "Two dollars from Mrs. Wade, Roy! And the bills are brand new! Oh, for Pete's sake, she didn't have to do that!"

They gave their windfall to the church youth group so the kids could go swimming, which had always been a huge treat for us as children. My dad's health continued to improve. Soon he could drive a tractor again, which worried all of us but made him very happy.

In fits and starts we got the picket fence painted its final coat of creamy yellow, to match the house. Mom and Dad came down one day to paint. Richard and I kept the brushes in plastic bags in the freezer so we could grab them and paint as we had time to do so and not worry about getting them clean each time.

My flower beds were starting to look tended, and the garden pumped out peas, lettuce, summer squash, beans, tomatoes, and whatever flowers I'd planted at the ends of the rows. We sweated to get the horses' hooves trimmed. I did the ones I could do alone, and if a horse had very tough hooves, Richard and I did it together. I held up the hoof, told him where to cut, and he muscled the nippers through.

Chapter 13

The weekend of the Suffolk Horse Association gathering arrived. The weather was better—less hot and humid. I spent the first morning, after distributing signs directing people to our farm and helping with various set-up jobs, driving Margaret, Lucinda, and Katrina—we called them the Dream Team—on our two-way plow. It was Margaret's first summer of work, and she'd been doing well. She matched her mother and aunt in height, endurance, and intelligence. It was a pleasure to work those three, and I was proud to show them off that day.

In the afternoon I hitched Stella and Jemima and let Jay, a long-time horse farmer from Vermont, rake hay with them. He showed me a line-management trick I had not learned yet, looping the line into the breeching and doing a sort of crochet stitch to keep it out of the way and yet quickly accessed. It would have been nice, I thought, if we could have had that kind of mentoring as we learned to use horses on the farm.

For two days we worked horses, talked horses, showed horses, met new Suffolk horse enthusiasts, enjoyed seeing and sometimes driving other Suffolk horses trailered there for the weekend, and in general were in a kind of Suffolk horse heaven. On the final day, Sunday, it rained. People packed up and left, vowing to stay in touch and to meet again next year, when the gathering would be in Ontario. The final guest drove away, and Richard and I nearly fell to our knees in exhaustion and gratitude—nothing bad had happened. With all those other horses around, all the visitors there to just take in the spectacle of several teams at work in the fields, all the ways someone could have gotten hurt—we hadn't even had to get out the band-aids. It was a very successful event.

When Peggy came to visit a week or so later, we were so far behind from event preparations and the event itself that I could not take time off to be with her. She had to work with me, which she willingly did.

"I'll be out in the barn when you wake up tomorrow," I told her one night before we toddled off to bed. "Just come on out when you feel like it."

"What will you be doing?" she asked, pausing for the answer before she shut the door to Marian's room, where she was sleeping. Marian had moved into our room for the time being and was happy in her mermaid sleeping bag on the floor beside our bed.

"Unloading hay, sad to say," I told her. "Richard has to go to work, and we need the wagon empty for when he comes home so we can bale. Don't feel like you have to help. I can do it."

"Hmmph," she said. And shut the door.

Next morning, as I began throwing bales from the top of the load over to the loft, Peggy called up to me from the ground beside the wagon, "Where are the gloves? I don't want to get my dainty hands all full of calluses." I directed her to a source, and she came back and clambered up to help. I flung the bales over to her and she stacked them. Or she slid them over on a board, once we were down lower, and I stacked them.

"Oh, a couple of hay honeys," Richard called up when he came out, nicely dressed for a day at the office.

"Come on and let us hug you." Peggy held out her sweaty, dusty arms and he backed up a bit, laughing.

"If you can rake that field once the dew is off, that would be great," he said as he headed for the car.

Marian and Peggy harnessed Jemima later that morning. It was hilarious to me, watching the two of them working together. Marian was far too short and not strong enough to lift the harness onto Jemima's back, but Peggy, with some explanations from Marian, could. Apparently Marian was not exactly clear, because at one point I saw the two of them pulling the hames forward from where Peggy had swung them over Jemima's backside, as if dressing her in pants.

They did it, though, and Peggy rode on the forecart beside me as I drove, and the hay rake flipped the windrow behind us. Before long, Peggy was driving, as I had expected she would want to, and that summer afternoon passed in the smell of drying grass, sweaty horses, and pleasant conversation well punctuated with laughter.

I missed her terribly when she left, but by then we had two big things coming up. Marian was going back to school, for one. Our little Montessori school class size had gone from six students to thirty-four. We'd all worked during the summer on making more of the Montessori teaching materials and in getting a modular building sited on a parcel of land we'd bought. Truman would start kindergarten, and I would be alone during the day for the first time in years. What would I do with myself?

Well, there was an acre of sorghum to help with that. Richard had found a sorghum press in Iowa, making a daylong trip down there, following the seller out into the countryside and down into a ravine to haul up this heavy, cast-iron set of rollers that would take in the sorghum cane, compress it, and shunt the juice down one side into a bucket. He arrived home well after dark, and in the morning I puzzled over the thing until he came out and explained how it would work.

"We'll mount a tire here for the flywheel, and then that belt I bought at that auction last year—it'll go from the tractor to the flywheel, and that will make the rollers turn. Except it has to be geared down, or the rollers will spin too fast. I have to figure that part out."

He pretty much forgot about me then as he pondered the machine, working out gear ratios in his head. Well, I was used to that, and very glad I was not going to have to do it myself. Getting the sorghum from the field down to the barn to pressed was looking like a major undertaking. The canes were about eight feet tall. The books said that it should be trimmed of its leaves before it was cut. One did this with a machete, stroking down from overhead, stripping the leaves from the corn-like plants on each side of the row and then going up the next row and getting the inside leaves stripped from those. We liked seeing the leaves on the ground as we moved along. What drew us to sorghum was that it could be returned to the land, once the juice was extracted. What we took from the soil would go back to it, mostly.

Slicing off the leaves was probably the easiest part of harvesting sorghum. Next, we used our machetes to hack the plants from their roots, piled them like slender green logs, and hefted them onto a hay rack. If we stopped to rest, our muscles seized. We chopped and lifted, chopped and lifted. The stalks were heavy with juice, which pleased us—but gosh, a bundle of them was hard to lift and carry.

We heard of someone nearby who had a horse-drawn corn binder. We went to look at it and to ask if we could borrow it. He agreed, on the condition that it never be left outside overnight, and we complied with that by rolling it past our car every night, putting the binder in the garage instead.

With the corn binder, I could drive Margaret, Diana, and Katrina along the row. The machine cut the stalks, they moved up and were held in place, then tied into a bundle, and dropped onto a platform. Once there were a few bundles on the platform, I kicked a lever and the platform inverted the bundles onto the ground in a single pile. Modern living!

***

The night before school started, we worked on the building and classroom until dark. Even so, the water and toilets didn't work, and the kids would have to use the facilities in the nearby building we'd purchased, which we were going to renovate over the winter. We were a weary group of moms and dads as we headed to our cars, kids in tow.

In the morning we stood together, watching the children proudly troop past us as they marched into the new building for their first day of school. One of the younger brothers, still a toddler, somehow got loose and joined the parade—only to be hooked back by his laughing mom just as he'd nearly made it to the classroom door.

Truman was nervous, but in the photo I took of him that morning, he holds his lunchbox and looks into the distance almost nobly, as if he sees a future none of us can imagine.

I stopped at the grocery store on the way home and kept patting at my hips and thighs, worried that I'd lost Truman somewhere. How odd it felt, not having him with me! "He's off to school now, huh?" one of the clerks said as he passed. "You'll feel funny for a while," said another, a woman who was stocking a shelf. I didn't know their names. But we passed each other all the time in the store and that is one way to get acquainted, I guess.

Richard got the sorghum press into running order. We took out our maple-syruping pan, set up the cement-block box for it to rest upon as the fire burned below, and assembled buckets to catch the juice we hoped to press from the stalks.

One day we hoped to have a horse-powered system, but for now Richard fired up the Allis-Chalmers and backed it to make the belt taut around the old-tire flywheel. He started the belt turning, fiddled with the situation for a bit, and shouted at me over the engine noise to start feeding stalks into the rollers.

I did, tentatively at first and then with more confidence as the rollers sucked them in and spat them out the back and juice flowed down a spout and into the bucket. Richard joined me to look at it—green, foamy. Later, when she saw it heating in the big pan, Marian would observe that it looked like horse drool, which was pretty accurate.

We made about fifty gallons of sorghum molasses. The farm smelled of woodsmoke and caramel. I fell asleep at night still feeling as if I were feeding stalks into the rollers. That is, when I COULD fall asleep. My anxiety was getting the best of me.

For years I'd been watching farm fields near the cities and around towns fill with houses. Sometimes, passing a development that seemed to have sprung up out of nowhere, Richard and I would ask one another, 'Where do all these people come from?' Deeper thought would have told me that if people are going to have large families, as my parents had, there are going to be a lot more people, and they are going to need homes. I did realize that. But it seemed wrong to be building on farmland. People need places to live, but they also need to eat.

I worried about it more and more, thinking about the future my children and their friends would face. I fretted about the increasing population of the earth. I imagined our small city becoming a boom town, as nearby Woodbury had, growing in vast sheets of roofs we could see from the freeway on our way to the city. I worried about the world getting warmer, the animals that would suffer as a result, and the pests that would invade our area, making it increasingly difficult to farm.

Richard, when I shared my fears with him, comforted me in characteristic fashion—which is to say, not much at all. "You're right," he said. "Overpopulation is the world's biggest problem. And climate change. You can do your little bit, but really, there's not much that will help. There's so much carbon already . . ." And then, seeing how dismayed I was, he tried harder to be of solace, telling me gently, encouragingly, "Populations really control themselves naturally, with plagues, and, um, ah, (here he was watching my incredulous face) I think we might, you know, survive."

Clearly I was going to have to find my own comfort if I was going to carry on with any kind of hope. As I had with the death of Superman, I put Richard's reality in one place and made up my own and kept it in another. In this case, though, I could actually take action. I found a book in the library called *Rural by Design*, by Randall Arendt. I learned that ours wasn't the only area being eaten up by subdivisions and that there were ways to incorporate housing with a rural landscape and continued farming. You just had to convince the people in charge to make ordinances that would direct growth to particular areas in particular ways.

In our local newspaper I saw a notice that the Land-Use Ad Hoc committee of our township would meet that evening at our town hall. I tucked my library book under my arm and walked into the old building, which had been built in 1907 and was used as a schoolhouse till 1960.

A few people were gathered around a table at the front of the room. I sat in the first chair I came to, an old school desk. Apparently it wasn't all that usual to have an audience for a Land-Use committee meeting, as all heads turned to look at me.

"Come on up here," one older gentleman called. "We've got the stove on. It's nicer up here."

It was a chilly evening, and being near a stove sounded good. I joined them at the table, introduced myself, and learned their names. The older farmer who'd invited me up there, Ed Hanson, noticed my book. "I've read that," he said. "Lots of good ideas in there."

I was surprised and encouraged. In fact, I left that evening feeling as if there was some small hope in the world. At least I was engaged in trying to save the landscape and way of life I loved and valued. And I wasn't alone.

***

We read about an auction to be held about an hour's drive away from us. It was to benefit an Amish family—the mother and father had been riding in their buggy one night when a semi-truck hit it, killing the mother and badly injuring the father. Eleven children, ages nine months to nineteen years, were left to manage on their own.

We loaded up some quarts of sorghum and a spring-tooth harrow to donate to the auction and drove over there. It was a cold day. The line of vehicles, tipped sideways into the ditches along the road leading to the small community where the auction was to be held, stretched for about half a mile on either side. The community—indeed, it seemed, the region—had turned out to help.

Individuals and organizations, churches and stores—everyone had donated something to be sold for the cause. The crowd milled around the tables filled with quilts, dishes, canned goods, clothing, shoes, and toys. Crates of chickens, ducks, and turkeys; a cage of kittens; sheep and goats tied along a fence; two sloe-eyed, darling Jersey calves; even a few horses—Marian and Truman were beyond fascinated. Occasionally a chicken would lay an egg—first a hen in this crate, then a hen in that one. Truman ran excitedly from one bird to another, trying to predict when the next egg would be produced.

"Mom!" Marian exclaimed. "The goat!" We watched as a wet, pink balloon protruded from the backside of a goat and then, slick as could be, out popped a baby goat. The nanny turned to lick it and then, to her surprise, out slooped another one. "For goodness sake!" she seemed to say, and she set about licking them and then letting them nurse. Marian and Truman were parked, agape, just feet away, and they watched adoringly as the kids dried off over the next couple of hours and began to look like small, furry collie pups.

I liked looking at them, too, but the smell of the sheep and goats was so rank that I couldn't stand there for long. I assisted an Amish woman in running down a rooster, and I found a pair of work boots in my size. They caught my eye because they were the same kind of boots my dad wore on the farm, and I'd been wearing men's size-ten work boots for the past years. That sounds worse than it really was, as I seldom needed work boots. Except for baling hay, I wore old running shoes on the farm. But now, here were some boots in my size, and with a winning bid of $47.50, they were mine.

We left the auction with the auctioneer's patter running in our heads: "What'llyougiveme, sixdollah, fie,fie, fie,—HEY!" All we bought was the boots, though we'd been lobbied hard on the mother goat and her babies.

***

"What did you do at school today?" I asked Truman as he and Marian climbed into the car.

"Work, work, work," he said complacently as he fastened his straps. He was enjoying school more than I'd expected. For one thing, he'd met another small boy his own age, Sam Cunningham. Truman referred to him always as Sam Cuttinham, and they were a pair. When the school organized a field trip to hike along the bluffs over the St. Croix River, their teacher specifically prevented them from being each other's buddies for the trip.

"You'll be absorbed in talking to each other and you'll walk right off a cliff," she warned.

We liked Sam's parents, which was a good thing. The boys went back and forth, Sam at our house, Truman at theirs, though we lived a good distance from each other.

One day I called to them as they played outside on the garage apron, "I'll be ready in just a minute to take you home, Sammer!"

I heard them stop what they were doing, and Sam sighed heavily. "Back to that old lonesome feeling again, huh, Truman?" he asked.

Because of the lack of structure in the classroom, we weren't sure what Marian was doing all day. At her conference, I asked her teacher that question, and she laughed. "Marian is the busiest of the kids. I think she needs a twelve-step program on relaxing," she said.

That wasn't too hard to believe. One afternoon I was reading to the kids and fell asleep. When I woke up, they were playing next to me, reading our encyclopedias and reference texts. Marian had completed a "book" (pages folded and stapled together, text heavily illustrated) about sea mammals, and Truman was trying to do one on seals. Marian then wrote one, with some help from me, on killer whales: *Killer whales or eckxulnt huntrs beacus thay wrck togither.* Still having some trouble with the spelling, I noted. From killer whales we went on to collaborate on a chapter book about big cats.

They had a collection of rocks with fossils in them, and Richard had found his old geology hammer and let them use it, with the stipulation that they wear lab goggles to protect their eyes while they worked. They used a set of small screwdrivers as picks, and often I came in to find them "fossiling," as they called it—crouched over their rocks and tools, looking at me though the fogged goggles, and talking as if they had terrible colds in their noses.

Because we tried to visit Richard's parents on the Oregon coast almost every year, both kids were interested in the ocean. Somehow I got hold of a roll of newsprint, and this led to a phase of trying to draw full-sized whales.

Our house couldn't accommodate anything larger than about twenty feet, so they compromised and made smaller-sized drawings that required a lot of tape and space to hang. Truman used up an entire blue crayon on his skinny blue whale, and that led to quite a fracas with his sister.

Their school put on a dinner play: Hamlet. They learned the story and made up their own lines. Marian's friend Jake, a calm, sweet-natured boy who sat patiently as the crown pressed down on his ears, played King Claudius. Truman's friend Eddy was Laertes. He got to wear a cape and took especial pleasure in ripping it off as he leapt into the sword fight. Marian and Truman were cast as Rosencrantz and Gildenstern, but when Truman found that his part involved neither cape nor sword fight, he refused to be in the play at all. Marian had to play both parts AND introduce the play, and she was fine with that.

Richard was gone for two weeks that January, and the kids were sick with coughs—Truman missed a week of school with his. Temperatures outside held lazily around freezing, with rain sometimes deciding to fall as snow. It was terrible weather for the horses, and I fed them extra hay. One windy morning I went out and, after stumbling through the knee-deep snow that floated on inches of water, found them at one of the three-sided shelters scattered around the pasture. They'd herded the foals into the shelter and gathered around it, keeping the young ones out of the weather. Sometimes I loved those horses for their beauty, sometimes for their willingness to work for us, and sometimes for the way they showed me that they were thinking, feeling beings—the way they almost let me, in a way, be part of their herd. That morning, when they looked over at me, their eyes told me they knew what they were doing and had things under control. Go get our hay, they said. We got this.

***

"I have an idea," Richard said as we drove home from picking him up at the airport. "I have to go to DC in a couple of weeks. How about if we all go? I have enough frequent flier miles piled up. And the company will be paying for my hotel."

"And the museums are free!" I was thrilled. For one thing, my friend Kim, whose sweet son Phillip was Truman's friend, had moved to the DC area the previous year. We could see her and Phillip, go to the Smithsonian, tour the monuments—it was going to be great.

And it was. We loved the suite in the nice hotel with the glass elevator on the OUTSIDE of the building. There was a pool—we'd brought swimsuits for that. Breakfast was free with the room, and kids ate free in the restaurant at night. For a couple of pennypinchers, this was a perfect vacation.

I'd only been to Washington to protest our country's involvement in El Salvador, and though I'd done my best to take in as much as I could during the few hours I had there, there was so much left to see. I was interested in every one of the Smithsonian museums, but our family's great interest is natural history, so that is where we started.

I had high hopes of covering a lot of ground, but once we were in the Hall of Bones, we stalled out. The kids raced from one skeleton to another, noticing the similarities between the whale's fin structure and the bat's wing structure. I pointed out how the horse's leg had once ended in toes, and now a couple of the bones were fused, sort of, to the sides of the cannon bone. Richard and I kept edging toward the door to another gallery, but at last we gave up.

At home, Marian and Truman kept a collection of bones on the seats of the bobsled in the barn. If you came upon it unexpectedly, you might think they were a couple of ghouls. The vertebrae and shoulder blades and skulls of long-dead dogs, deer, probably skunks—who knows what-all—were lined up, each in its own category. And now they were in this glorious palace of bones. We were there all morning.

Later we went to the Tomb of the Unknown Soldier and watched the changing of the guard. In a way, the pomp bothered me. Yes, it's good to be reminded of the sacrifice our soldiers have made. But the streets of the city were full of homeless people, many of them carrying signs that identified themselves as veterans. Wouldn't it be better to spend our money on these individuals who served our country and did NOT die? (Marian had already asked if they came to our house, would we let them stay there?)

Just then, as one of the guards clicked his heels and turned smartly, and I was thinking these thoughts about how veterans could be better honored, Truman leaned over and whispered, "What about if the guard falls asleep in the night and someone gets in to hurt the dead guys?" This surreal scenario seemed to fit what I was thinking.

***

Back at home, where our vet and neighbor, Peter, had been in charge—as we were at his place whenever he and his family traveled—we were beginning to watch for signs of imminent birth. We had nine mares we expected to foal.

Early one morning I was awake, lying in bed, fretting about something or other, as is my wont at that hour—it was 3:30, I saw when I checked my watch. I heard the horses galloping down the farm road and knew they had gotten out yet again. We had a hayrack full of sorghum stalks we hadn't been able to spread back onto the field yet, and the horses loved snacking on them. Horses have such finicky digestive systems—we were not willing to risk letting them colic on sorghum cane.

We got up and went out to open the gate and let them back into the pasture. I noticed that Inga, our sweet little Fjord mare, was not with them. We had to go back up the hill to work on the fence, and as we walked, we watched for her. She was the usual Fjord color, a light dun, so she would not stand out against the snow as one of the Suffolks would. Toward the top of the road we saw her in the moonlight. And something beside her. She'd had her foal, and he was already up and looking for breakfast.

It was a frosty morning, and once the sun came up he looked like a little horse elf, so furry and dark eyed—so Nordic. I asked our friend Trygve, who had grown up in Norway, how to say frost in Norwegian. "Reem," he said. "You spell it R-I-M, though." Ha, I thought. Kind of like our English word "rime" for frost. We named the little colt Rim, pronouncing it "Reem." I rubbed Inga's furry ears and her itchy spot under her forelock, and Rim explored my jacket with his wee, cream-colored nose and lips. I'd hoped for a filly from Inga, but every horse born healthy on our farm was a gift. The ghosts of Lucky Romper and Superman hovered over every birth.

**

Early Spring Farm Walk

Autumn speaks through the spring—-see
the garlic sticking an exploratory finger
through straw as if testing the air,
and this field where the turnips
lie as if hailed onto the gummy dirt,
squeezed out of the soil by frost, such
a cover crop as deer dream of—and me,
having loved all winter watching them
against the hill.
Rye coming up in this field, good
to see it so green and spreading, planted
after the beans came off, and now
the prairie's dry rustle of husks and brown grass
that will be burned this year, flames leaping up
just as this couple of mallards jumps from the pond

as I pass.
I love this trail on the ridge's broad spine
where the passing winds lift me
And yet lace me down,
where I pass a bare, raw spot
—we buried the yearling colt
in November after his baffling illness
and suffering ended, gave him
to this place we love and love more now
for his place in its future, his
fine, patient heart
beating strong through the land.

***

"I want to go spend some time with Bob Erickson," I said to Richard one day as we sat back in our chairs after dinner. The kids were upstairs, and we were enjoying a bit of quiet time together. "He's got such a good reputation as a trainer. I'd like to see what his tricks are and how he does things."

"Sounds like a good idea," Richard agreed. "We've sure got a lot of foals that are going to need to be trained." He stopped and cocked an ear toward the stairway. From Truman's room we could hear the kids singing a song I'd been singing to them as a lullaby all their lives. Very sweetly they sang, Hush little baby, don't say a word—and then, just as sweetly, because who needs to know all the words exactly anyways—doink doink doink-doink, doink doink doink take a breath, doink doink doink doink, doink doink doink.

I wrote to Bob Erickson and asked if I could come and work with him one day, and he called to invite me for the following Saturday. It was a three-hour drive to his farm, and I arrived feeling rather stiff from sitting. And cold, once I'd left the car. We'd had warm weather the previous day, and I very foolishly hadn't checked the forecast. This was to be a very cool day, and I was not dressed for it.

"You ready to go?" asked the big, eager-looking man who answered the door. I said I was, and we were off. He showed me how he harnessed his team of Fjord horses and then made them stand as we went to look at the rest of his herd. This was the third time the team had been harnessed and hitched, he told me. They already stood well, which I had found to be the most difficult thing to teach a team—mostly because I was impatient to get going and rarely made them wait to step off. Bob was patient and exacting. He had each horse fitted with a rope that encircled one of its front hooves and went up to the hame ring, then back to where he sat in the cart. If the horse became restless and lifted that hoof, intending to start walking, he drew the rope back, holding the hoof up. The horse was immobilized, but gently. It could not get its foot back till it stopped struggling.

Bob was a master of timing, and each time a horse fidgeted, he took control of its hoof. Once he had the behavior he wanted, the horse got its foot back. He let me handle the ropes for a while, and I saw how important it was to pay close attention to body language— such as the movement of the horses' ears. In trying to be johnny-on-the-spot with the rope, I jerked it and scared the horse, doing no good at all.

We spent all day with the horses, driving them on the road and practicing turns, making sure to expose them to traffic of all sorts. I tried to keep my shivering under control as we drove into the raw wind—oh, great, now trotting, so the wind was even stronger—and I fantasized about how hot I'd get the car for the return trip. Bob clearly loved being out with his horses, and it was inspiring to be with him. I'd come there to learn his tricks, but when I finally got into the car and cranked up the heat till my lungs almost burned as I breathed the air in it, what I realized was that his most important trick was time. The time he spent with the horses was the reason his teams were so quiet, reliable, and enjoyable to drive. It was an excellent lesson for me.

Back at home, nice and warm from my overheated car, I saw that Dinah was ready to foal, and I spent that night in the barn, cold again. And, in true Dinah form, she did not foal until the next day, later in the afternoon, when I was in the house.

Katrina and Jemima were bagging up. Katrina even had a bit of wax at the end of her nipple, so she was closer to foaling than Jemima, I figured. But there was Jemima, late one afternoon in April, with a filly at her side. The baby had the largest blaze we'd ever seen on a horse. Marian wanted to name her Blaze. "Don't call attention to it," Richard and I said at the same time. She reconsidered and decided to name the filly Ellen, after her teacher.

This would be Katrina's fifth foal. Lucky Romper, Carl, Superman, Kiss-Me, and now this one. Certainly it was time, I thought, for her to have a filly. I wanted her kindness, good sense, and beauty passed along to a daughter.

This might be the night, I thought, and I set up a lawn chair under the shop light and read Annie Proulx's *The Shipping News* through the evening while Katrina munched on hay in the foaling stall. No foal. At bedtime I climbed to the loft, where I had a cot set up in a place that allowed me to check on a possibly foaling mare without leaving my sleeping bag. I really had it down to a science.

Katrina woke me with a racket she was making by banging on the gate.

I saw that her back was wet. I dozed until I heard her water break and then watched as she labored and strained to have her baby. I finally went down and watched as the nose and hooves protruded. She worked and worked. I grabbed the foal's legs above the fetlock, as Peter had shown me how to do, and pulled toward her hocks when she was straining.

Finally the foal's shoulders escaped the birth canal and I had a lapful of struggling wet horsehide wrapped over bones. The foal's scrawny body was sprawled in such a way that I could see she was a filly. A filly! I had the name picked out already. After such a long wait for a filly, I thought the name "Patience" would fit perfectly, which it did—for reasons other than that, which I would learn later.

I was so happy that I sat in the straw, off in a corner out of the way, as Patience and Katrina got to know one another. The usual baffled struggle to figure out how these legs worked resulted in the filly's standing up. I was astonished. Granted, I'd been waiting for this filly for years. I loved her mother. I would have been happy with a healthy colt; I would have been happy with a regular horse. But this filly, not an hour old, was already the most beautiful horse I'd ever seen. There was something about her that stood out, even as the birth fluids dried off her furry baby coat. I didn't want to leave the barn. Usually I went back to bed in the house once things were in order with a night birth in the barn. But instead, this night, I climbed back up to the loft and into the sleeping bag, and I turned toward the stall so the last thing I'd see as I fell asleep would be Patience, the beautiful filly for whom we had waited so long.

It wasn't just me. When Richard came out in the morning before work, he expected to find that nothing had happened. I was just climbing down the ladder. "No foal?" he asked.

"Look," I said.

He turned to the stall, where Katrina was standing over her sleeping baby.

"Oh, good," he said. "I thought you'd come back in once it was born. Something wrong with it?"

"Just wait," I said, because Patience was beginning to stir. She got her legs organized and rose from the straw.

Richard watched her stretch and walk over to Katrina for a drink. "Wow," he said. "That's a fancy horse."

She wasn't tucked under, as many foals are at first. She was balanced and rounded, pleasing to the eye. I often had to stop myself from giving foals silly names because they looked so homely and no-account. This filly bore her name easily. She was worth the wait.

Foals popped out of mares every few days for the next few weeks, it seemed. Diana had Christina, Margaret had Fiona, Stella had Jeff, Dinah had Clara, Rosalie had Ramona. We expected Lucinda to foal later in summer.

Chapter 14

Now we had more than thirty horses. We kept them in various paddocks around the farm. I maintained a For Sale list and tried to make it widely known that we had HORSES FOR SALE. But we were not selling them at the rate we'd hoped. Nonetheless, Richard was already talking about getting the mares bred for next year. I balked. I argued and fumed. We had too many horses. We couldn't spend enough time with each one; we could barely catch some of them. We weren't doing a good job.

It was a Saturday morning. I had several things I wanted to do that day, once we had the horses wormed. We went out with our boxes of worm paste, our halters, our buckets of wheat (we used our wheat for bread AND to catch horses).

We did the easy horses first. The mares, the stallions, the gentle geldings. But the young stock did not want to be caught. We'd found that once the foals were on pasture with the herd, they became wary of us. For their first two years, they did not want to be caught or handled. And then, almost as if a page had turned, they would start coming up to us to be scratched. It was easy to put a halter on them. They let us pick up their feet, after a fashion. They were respectful of our space.

But those couple of years of wildness made it an awful chore to handle them for worming, vaccinations, or any other reason.

On this particular morning, I was thinking about the jobs I had waiting for me once we'd finished this one. We finished worming Stella, the last of the easy ones, and set about catching Elaine. Half an hour later, we still had not caught her, and there were another six or eight to go. Not to mention the wrestling match we had ahead once we did catch each of those wild horses.

Suddenly I'd had enough. I was sick of the farm, sick of Richard, sick of work. "I'm sick of this," I said to Richard. "We have too many horses to take care of them. I'm done." And I walked away. Back in the house, put some underwear and my toothbrush in a bag and I drove away.

I headed for Inga's house. I wasn't leaving Richard. I was just so frustrated about how things were going and that I could not get through to him that it was important to me to do a good job with the horses—to have fewer of them so we could take more time with each one, get to know them better, and enjoy them more. He saw that this breed of horse was rare and felt the importance of getting the numbers up. We were both right. But his version of right was killing me.

I called Inga from a pay phone in Durand. "Hi, are you home this weekend? Would you mind if I came for a visit?"

She sounded surprised but said, "Of course. When should I expect you?"

Oh, the gift of good friends.

When I moved to Illinois after leaving my job at the newspaper, and my family, and my great friend Peggy, it was to a small town on the Illinois River. I had a bicycle for transportation, a sleeping bag for a bed, and a dresser and a vanity that I used as a desk. I didn't mind the lack of furnishings or a car at all. I missed Peggy. I missed having a friend to go to the movies with, to talk about whatever book I was reading and learn about what she was reading, to go swimming or running with, to tell about whatever funny thing had happened during the day. I was lonely.

I filled my off hours with what had always helped—exercise and nature. I rode my bike to the state park and hiked in the canyons that branched down to the Illinois River. I rode out into the countryside and began to fall in love with flat land and huge sky. I wrote to Peggy and other friends, typing into the night at my desk in the rounded corner of my apartment in a formerly grand home. Without the cinnamon-sugar flour tortillas and the M&Ms, I began to lose weight. I ran. Sometimes Peggy and I splurged on long-distance phone calls.

Once in a while, my boss and her husband had me feed their cat and carry the mail into their house when they were traveling. Sometimes there was an envelope or a package in their mailbox that had come from Norway. Their daughter Inga lived there lived there with her boyfriend. I met her when she came home for Christmas. Then, a month or so later, she came back for good, sans the fellow and lonely, too, for a friend. She liked movies and books, running and swimming. She worked at a separate part of the company, and we often sent notes through the interoffice mail. "Scotty's Tacos tonight?" And then I'd be waiting on the street for her to come by and pick me up after work and we'd hit Scotty's and laugh about what had happened that day.

Again, luckily for me, Inga was also a writer of letters. When I left the magazine, she and I corresponded often, talked on the phone, and visited each other whenever we could. Later on, she worked in Germany for a while, and I took Truman with me to visit her when he was seven months old. Just as Peggy had in Oregon, Inga drove me around parts of Germany and the Netherlands to look at horses. We both cried when I left.

Now she lived in the Illinois Valley again, had her own house, was an executive at her family's company. I barrelled through the distance between us and arrived after six hours, so relieved to be in a different place, with my good friend, clearer in my mind after the hours of thinking on the trip.

I called Richard that evening to explain that I would be home late the next day. I just needed some time. He said he had been thinking, too. He agreed—we had too many horses. We'd work it out when I got home.

In the meantime, Inga was, if not thrilled with my company, at least putting on a good show of it. She didn't ask questions, just offered food and drink and the possibility of a movie on the VCR. We walked on the tow path beside the old canal and we ate potato chips and sour cream as the rom-com played out on the TV screen. I finally told her the reason I was there. Her beautiful face said how sorry she was, even before her voice did.

"We'll work it out," I assured her as I threw my little bag into my car and slid in. All around me the Illinois springtime was shouting via flowers and various shades of enlivening green. She stood on the sidewalk and waved to me, and I waved to her, until I turned the corner and we could no longer see one another.

At home, once we had the kids in bed and could talk, Richard and I sat down together at the table. "I've been thinking about it," he said. "You're right. We have too many horses. They just aren't selling. At least, the ones we have for sale. The young ones." I started to see where this was going. "We're going to have to sell the older ones, the ones that will sell more easily. People want broke teams, and we have them."

He was right and I knew it. But . . . I loved the horses, the ones I'd spent so much time in the field with, who had worked and worked for us and made it possible for us to stay on the farm and have this life.

"I talked to Ron," he said. "He wants to buy Lucinda and Katrina, once their foals are old enough to wean."

Katrina? Not Katrina.

"And Stella and Jemima. Not right away, but eventually. I think we should keep Diana so we get more foals from her. No one is going to buy Dinah, so we'll keep her. Once you've got Rosalie going better, we'll put her up for sale."

Stella? It had been hard enough to part with Jane. Katrina? I began to understand that to achieve my goal of having fewer horses, and Richard's (and, to be honest, my) goal of increasing Suffolk horse numbers overall, I was going to have to sell my friends.

***

My parents came down for Truman's sixth birthday. They brought his cousins, Roy and Carter, who were also six years old. The wait for them that morning was excruciating for our boy. He helped me do chores that morning, rode the big trike up and down the road, rode his bicycle up and down the road, and finally, at about nine o'clock, exclaimed in exasperation, "They SAID they'd be here around lunchtime!"

Finally the blue Dodge nosed up the driveway and my mom parked the car, from which spilled two eager boys to be met by our eager boy, and that was all we saw of them for a while as they hurried off to whatever adventures they might find.

My dad was much slower about leaving the car. Mom gave me a significant glance as she watched him unfold and stand, steadying himself against the vehicle. He was not as strong as he'd been when I last saw him. She was worried. That's what she said with her look.

I have a picture of him from that day. He's at our kitchen table, loose and happy, noodling with his grandsons. That's the last time I saw him like that. After the visit to the doctor, after my sister called, swallowing her tears, he seemed to feel as if he took up too much space in the world. He lost even more weight, he no longer sprawled and joked. He waited politely for the cancer, this time inoperable and untreatable, to have its way.

***

Peter, our friend and vet, stopped by one evening on his way home. "Say, would Marian be interested in keeping a pony for the summer? She can ride him. He's a nice little guy. He just needs to be put in at night. If you keep him on pasture all the time, he'll get too fat and he'll founder. He's already got a little bit of that in his feet."

When asked, Marian made no bones about saying YES. We went to visit the pony, who was stabled near town in a barn that would soon be torn down so the land on which it was built could be developed. The woman and her daughter who owned the pony wanted to keep him. "He helped Jenny through the divorce," Kathy, the mom, told me. "We'll never sell him. But Jenny's too big to ride him now, and I hate to have him just sitting around. He likes kids." She showed us how the pony, Spooner, responded when she crinkled a bit of cellophane candy wrapper. His head whipped around and he trotted over to where we stood at the fence. Jenny produced a red-and-white striped mint and he lipped it out of her palm and crunched it enthusiastically. We laughed, and they let Marian feed him one. We'd have to give him back after the summer, but for the next few months, we had a pony.

Marian could saddle him herself and bring him to me to make the last tightening of the girth. She rode off, and I could sometimes hear the snatches of chatter as she talked to Spooner about this and that. Truman didn't ride him but instead enjoyed standing up on his back, balanced like a circus performer.

"When she's sad or upset," my dad told me when I described Spooner and his life with the kids, "she'll go off with her horse. It'll be good for her."

I remembered how I used to get myself through milking in our dark, low-ceilinged barn, scared of my dad's unpredictable anger, urgent to get away—I'd imagine my horse in her pasture, going out to her in the dark, just standing next to her and breathing in that smell. My horse got me away from the farm and out into the neighborhood, where the people I saw raised their arms in greeting from where they hung laundry or drove the tractor. Marian and Roy's girl, out on her horse. Again. I always came home happier. And Dad must have noticed.

***

We planted a larger crop of potatoes. We decided against growing sorghum again—we were having trouble selling what we had, and the work involved in making it had been too much, at exactly the same time we had to get the potatoes dug and washed and stored and marketed. Our canola crop had been promising for a while, but one strong rain at the wrong time had pummeled it into the ground. We decided not to try canola again. Rye, spelt, wheat, buckwheat—we'd had success growing those. Even with the All-Crop, we had trouble harvesting. But we kept thinking that was a nut we'd eventually crack to our satisfaction.

Our beloved cats killed mainly rodents. Once in a while, though, to our distress, they killed birds. The kids and I had been visiting my parents. When we got home, Richard told us he'd found the two adult bluebirds dead near the base of their birdhouse. He could hear chirping inside. Marian was galvanized—Truman as well—and they resolved to raise the babies themselves. They brought them into the garage, settled the nest into a cardboard apple box, and got busy.

Marian pulled the bird books off our shelf and learned what bluebirds eat and feed their young, and then she and Truman set about finding insects. I had little hope the tiny birds would survive, but my children were deeply involved in seeing that they would. I called a wildlife vet who agreed with what the kids were doing and recommended that they include a few drops of water with the solid food.

Richard and I stood back and were pleasantly surprised that the chicks stopped peeping and slept, awoke, peeped for food—eagerly fed to them by our delighted and assiduous children—and then slept again. This went on for a couple of days.

Lynn's daughter Jayne came for a day while Lynn and her older daughter Ellen had an outing. She happily joined in the chick rescue. It was looking good for the bluebirds, we thought, when suddenly a howl erupted from the garage. We ran to see what was wrong and found Marian standing over the bird box in the garage, weeping inconsolably. "Diamond (a stray cat who'd been hanging around) ate them! She was in the box when I came out!" Truman joined her in grief. Jayne stood, looking stricken.

Even now I feel it—that sense of having failed my children by being too busy, not thinking things through, taking for granted that things would work out. I should have made sure they had the chicks secured in the cat carrier, or let them bring them indoors. Marian climbed the stairs to her room, shut the door, and wept for a while. Truman and Jayne sat miserably.

Then Marian came down for some hugging from me, after which she told Truman and Jayne that they should go check on the chicks in the other bird house. They trooped out, still in mourning, but arrived back in better spirits. When Marian had reached into the bird house, she'd felt the mother bird!

So they got the bird book and sat down together to study the pictures and determined it to have been a tree swallow. Then they collected a clip board and pen to go with the bird book and went back out to check on whether they were right or not. Later in the day, Marian officiated at the funeral for the bluebird chicks. Diamond was on thin ice for a while, but eventually they forgave her.

***

When I could, I drove up with the kids to see my parents. Each time Dad was thinner. He loved hearing about the farm and the horses. Our misadventures with the baler, the antics of the foals, my work with the town committee to preserve farmland—he liked hearing about it.

It was when we baled hay that I cried. My new boots from the Amish auction were the same brand and style that he'd worn all my life. Coated with hay dust, they did not look new. My feet looked as his had. I was doing what he'd done, stacking hay as it came from the baler. He'd always used a sideways motion of his foot to move hay bits and dust off the rack before putting a bale down on the wood. Now I knew why—he (and now I) wanted the hay to grip the wood and not slide around on the dust. I had a specific pattern for stacking bales, and I could stack a wagon of good hay bales tightly enough that I had no qualms about letting the kids ride it back down to the barn. I learned a lot of that from him, without knowing it at the time.

In the noise and sweat and smells of haying, I could cry. Richard was far enough away, driving the tractor. I wore sunglasses, and my tears looked like sweat streaks. It wasn't that I was embarrassed to cry in front of him, or that he didn't understand. I just didn't want him to feel bad, too, as I knew he would if he could not comfort me. No one could. My dad was dying—it was unthinkable. But I thought about it all the time, and on the hayrack I recited Dylan Thomas's poem about raging at the dying of the light, and Robert Hayden's poem about how his father got up early in the blue-black cold—as mine had—about the chronic angers of their house—as had been in mine—and love's lonely offices, and I saw my boot/his boot scraping the hay dust off the wood, and I raged against the dying of the light. I saved up things to tell him about when I saw him next.

We were fighting the potato bugs tooth and nail that late June and early July. I walked the rows with a backpack sprayer, wanding a mist of Bt onto the plants. One of the kids walked along the rows with a duster, wafting clouds of diatomaceous earth onto the plants, on the theory that the soft bellies of the larvae would catch and tear on the sharp edges of those tiny, ancient shells. Whoever was not dusting was simply picking adults off the plants. Such hated work! After a set number of rows, they switched jobs and continued in our battle against the bugs.

Dad enjoyed hearing about that. He'd told me, back when he was lying in the hospital after his accident, that "I used to feel sorry about how hard I made you kids work. But Mother and I tried to put some fun in there, too."

Kids should work. That had been his and my mom's philosophy, and now it was mine. Mostly out of necessity—we all had to pitch in. I tried to do as my parents had, adding in some fun whenever possible. One favorite activity was going to the south fork of the river that ran through the local UW campus. There's a bridge just above a little riffle, and I could spread a blanket in the shade, bring a book, and enjoy a quiet interlude as the kids played in the running water. Sometimes we walked up or down the narrow river. And we'd walk across campus to the gas station to get an ice-cream cone to eat as we walked back to play some more in the water. It was a nice break from our hot work on the farm.

A woman at our UU Society asked me if I would let her daughter come out and help us on the farm. She loved to work outside, said the woman. She needed something to do.

Cathy arrived one morning as arranged. She carried a small black dog with her. They had the same eyes, I thought, large and brown. Cathy was thin, about my age, and tense. She was so thin that I wondered if she'd be able to do much, but when I showed her a row in the garden that needed weeding, she set to work there as I began on another.

"Look!" she called to me after some time had passed. She stood, holding up her bony arm and waving an enormous dandelion plant and its carrot of a root. "Now, there's a weed!" she crowed.

I laughed. I was used to my kids being somewhat resentful about being made to weed. Here was a grown-up volunteer who was having a good time helping me. We were a match made in heaven.

After that, Cathy came out and spent a few hours with us a couple of mornings each week. She was on disability and couldn't be paid. I sent vegetables home with her—she never ate them, that I could tell, but she shared them with her parents, who were appreciative. She helped in the garden and she picked potato bugs. The kids grew to love her—first because she helped us, and therefore they had less work to do, but then because she was always nice to them. When I was working the horses in some field over the ridge, she and the kids would be assigned rows in the garden to weed or potato plants to de-bug. She had a small black dog named Misty Rose, and sometimes she brought her along, to the delight of Marian and Truman.

***

"Dad wants to come and see you and the kids and the farm," my mom said on the phone one day. "Harold and Leone will bring us down. Will Wednesday work?"

I had a couple of days to ready the farm. Cathy helped with the garden. I wore out a couple of teams hilling and cultivating potatoes. I mowed, weeded, mulched. We all worked to make our place look as perfect as possible for Dad and Mom. I knew they would appreciate it, and I wanted to make them proud, especially as they would have Harold and Leone, longtime friends — almost like parents to me and my siblings — with them.

Dad needed help to get out of the car. He stood and surveyed the yard, and I could see the approval in his face. My mom and Harold helped get him into a lawn chair, and I showed them and Leone the raspberries, currants, apple trees, the garden, the roses. We got Dad inside and the lunch I'd planned and made so carefully went over as well as I'd hoped. He loved ice cream, and we'd always enjoyed homemade ice cream at holidays, which we'd often spent with Harold and Leone and their kids. A few years before, Richard had given me an electric ice-cream maker, and I'd pressed it into service for this occasion, making the familiar treat and offering it under blueberries, another of his favorite foods.

He couldn't eat much, but what he ate he enjoyed. He needed a nap after lunch, and while he slept I took Harold and Leone for cart rides behind Inga, the Fjord filly—now mare—I'd named for my good friend. I showed them Harold the Fjord horse, named for Harold the man. I took a picture of our dear family friend with his arm draped over his namesake's neck.

Dad woke up and it was time for them to go home. He requested one last thing—to see Katrina's filly. I'd spoken of her often, and he wanted to see her. Richard helped me get him into the pickup and we drove out into the pasture, where the herd clustered around. We had nine foals, all of them beautiful and perky, but he noticed Patience and agreed—she was special. Then he wanted to see the crops. I drove up to show him the potato and grain fields, all of them looking their best. Within weeks we'd be unable to control the weeds in the grain fields, but on that day in mid-July, we looked like really good farmers. It was great.

Back at the house, he and Mom sat on a bench in the ell made by the picket fence and the house. They were backlit by hollyhocks as Marian rode Spooner across the lawn, so proud to show off her horse. I could have broken then, watching them watch my daughter come to them on that fat little black pony. Their faces said that they saw Marian's pride in her ability to ride this silly little horse, and they loved her for it. That they remembered me riding my own horse. That they appreciated the pony for being kind enough to let their granddaughter ride him safely. My parents were never well-to-do in money, but they took riches from the world, and found in apple blossoms, the fat thighs of babies, and dimpling little girls on stout Shetlands a wealth that billionaires can't match.

"I love you," I said to Dad as he settled back in the seat beside Harold for the ride home. It wasn't something we said. We had a history, as I have intimated. But time had eased the memories and reinforced the facts—he loved me. I loved him. "I love you, too," he said. Harold nodded at me and backed up the car.

Chapter 15

Richard, anticipating the various grains we would harvest later that summer, began designing and building a granary. Truman joined enthusiastically in the hammering. As Richard and I worked on the sides and rafters, Truman pounded busily on scraps of wood on the floor. One day we knocked off and set about picking up the wood scraps, only to find all of them nailed together in one giant and chaotic sheet of oddly cut two-by-fours and bits of plywood.

I began to wonder what Truman would do if we were to move to town and live in a house on a lot with just a yard and no barn, no tools, no animals. He loved riding his bike, and he constructed ramps with every board Richard could spare—not that he couldn't have done that in town. But other things maybe not so much.

One day Margaret had a limp. I couldn't locate the source of the problem. I trimmed her as close as I dared, but she had a split in her hoof—maybe that was it. I asked our neighbor, a master farrier instructor, if he would look at her. Graciously, he did.

He approved, to my huge relief, of my trimming job. And he treated the split by screwing a copper band-aid-shaped strip across it, mechanically holding the hoof together so it could grow out normally.

"Where do you get those strips?" I asked from where I stood at Margaret's head. Truman stood as close as Ray allowed, watching what he was doing.

"Copper pipe," came the muffled reply as he worked. "I cut the pipe with a saw. Then I snip it apart with tin snips. Then I pound it flat and cut it to the right shape. I use trophy screws to screw it in, but this mare has a thick enough hoof wall that I could have used something more substantial." He stood up, glad to stretch out his back.

A few days later, I found Truman hard at work at the anvil, ball-peen hammer in one hand and a wad of copper in the other. "What are you making?" I asked.

"Thimbles," he said. "For you and Marian." He'd been listening to Ray, too, and followed the instructions—he used a vise and hacksaw to cut a copper pipe he found near Richard's work bench, snipped it apart, and was now using a punch to form it into a thimble. He was six years old.

Rain threatened later in the day. We had hay down and ready to bale. The baler behaved and we got one rack filled. It was early afternoon and we hadn't seen the kids for a while, so we hauled it back to the yard and went in for a quick lunch, then headed back out. "We have another load to bale," I told Marian. "You guys can stay here and play or come up with us."

She chose staying near the house and playing—they were building an obstacle course for the cats, who, I am sure, were quite thrilled about it.

Richard and I went back up and had the hayrack about half filled when the shear pin on the baler broke. The brakes didn't work on the tractor, and we were headed downhill toward the woods and a deep ravine, so he cranked the front wheels at a sharp angle and remained on the tractor for a short while to be sure it was going nowhere. Then he jumped down, rummaged in the tool tray for the kind of bolt he used for replacement shear pins, and came back to minister to the baler.

I jumped down to help him. Suddenly the tractor, which had been rocking back and forth with the motion of the baler's packing arm, began to move. The front wheels worked out of their cranked position. I leapt to grab a bale from the load and fling it in front of a wheel of the hay rack, but it was too late.

We watched in dismay as the parade took place in front of us. Allis the tractor gathered speed as she trundled toward the woods, hauling behind her the baler, which towed the half-filled hay rack. We winced hard as the tractor hit the woods. The trees and bushes cracked under its wheels, and it was slowed somewhat, but not enough—there was too much pressure coming from behind. It could not stop, but went over the lip of the ravine and down the side, followed by the baler and the hay rack.

We ran. The tractor continued to strain but was now blocked by a tree too big to break or bend. Everything was stopped. Richard turned the tractor off and we surveyed the damage. Dollar signs popped up in my head—how much would it cost to replace all this equipment?

"How are we going to get this out?" I asked Richard. He didn't answer. He was already bushwhacking his way to the back of the tractor, where he unhitched the baler. Then he got on the tractor and started it. He backed up, getting free of the tree that was holding Allis back. Then forward, cracking down the bushes and young trees in the way. Then back, getting momentum to crack the next impediments — all at a dangerous slant that could easily have resulted in a rollover. Back and forth he rolled the big wheels, gradually working his way up and out of the ravine and then, finally, out of the woods.

Then we unhooked the hay rack from the baler, and we pulled the rack backward up and out of the woods. The baler was in too deep and hard for us to get just then. Rain clouds were lowering, and we could feel the occasional fat drop on our arms and hats. We busied ourselves grabbing up the scattered hay bales, stacking them haphazardly on the wagon, and then limping back down to the house as the sky disappeared, replaced by sheets of rain.

I thought of the load waiting in the yard. We hadn't put it in the barn, and it was now being ruined by the rain. The whole day had been an exercise in waste. We'd have been better off just staying in bed or going to some damn mall and eating French fries.

Richard could only push the wounded tractor so fast. All around me the bales were being soaked, as were the bales on the other wagon. I looked ahead to see how much farther we had to go, and I saw Richard motioning for me to look at something. Through the cataract coming off the brim of my poor, soggy hat, I could see the other wagon—covered with a tarp. The kids had seen what was going to happen, gone to the barn and found a tarp, and then clambered up and around the load, spreading it over the hay to protect it from the rain. That wasn't an easy job even for Richard and me, and I looked at it, frozen, as I pictured the process they'd had to enact.

When we finally got into the house and were met by two children with rather a lot of hay in their hair, we thanked them sincerely. The day had not been wasted after all.

Later on, Marian and Truman had a spat, and I comforted Truman by offering to take him up to see the crashed baler, which heartened him considerably. Richard was, to my astonishment, not only eventually able to extract the baler from the ravine, but also to fix it for the cost of one new part—seventy dollars. I was so impressed that I wrote a silly poem in cowboy style and sold it to a magazine for a princely sum of almost that much. So, as the title of the poem suggested, that was a Heckuva Day.

***

One late morning in August I was mowing hay on the south forty. I stopped to rest the horses and heard a plane passing overhead. The engine cut out, and I looked up in concern. The plane glided, then the engine started. The little plane climbed sharply and did a loop-de-loop. Then a barrel roll. Then some kind of maneuver in which the engine was not running and the plane fell, swish-swish from side to side, until I was afraid for the pilot and then at last the engine started and the plane swooped up to do more tricks in the sky. From my seat on the mower I had my own private aerobatics show, and I thought, "Oh, I have to tell Dad about this."

He hadn't left the house since returning from his trip to our farm, my mom told me on the phone. He was weak, and they'd brought in a hospital bed for him. The kids and I picked some chokecherries, one of Dad's favorite fruits, and drove up to see him. For all that day people came and went. Dad received them in his hospital bed in the guest room, able to talk in a low almost-whisper, smiling when Truman and his cousins Roy and Carter whizzed around his bed and shot out again, chewing on a chokecherry appreciatively, thanking everyone who stopped to see him. It was almost a festive atmosphere, what with the neighbors and nieces and nephews—even some cousins from out of town. Everyone had a quiet visit with Dad and then stood around outside saying how sorry they were that he was so near death. Then, because life was continuing for the rest of us and we were happy to see each other, conversation turned to other topics, as it does, and the sadness of the day receded for a while.

The next day Dad was not awake much. He lay quietly, and if people came to see him, he might flutter his eyelids, as if he wished he could open his eyes to see them but he could not. It was a close, hot day that turned into a humid night, and I spread a sheet on the floor beside his bed, as if he were a mare ready to foal. Only this time I would be of no service to him. No one could help him along on his journey. All we could hope was that he would have a gentle end and that he was aware of how much we loved him.

"Go sleep in your bed," my sister whispered. "I'll sit with him now. I'll call you if anything happens."

It was midnight and I got up stiffly, slipped in with Truman downstairs in the guest room, and slept until I awoke, sharply, and got up. Andrea met me at the top of the stairs. "He just died," she said. "Just now."

My mom stood at Dad's head. She'd been expecting this, as we all had, but expecting is not the same as experiencing. It seemed as if she wanted to do something but couldn't think what, and finally she leaned down and kissed his forehead.

I helped Craig from the funeral home fit Dad's feet into the body bag. Andrea helped with his head and arms. The hearse took him away from his house. Two or three of my brothers and their wives, Andrea, me, my mom— we watched the taillights disappear. Dad was gone.

It was such beautiful weather on the day we buried him. Perfect for baling hay, we said to one another. I gave the eulogy at his funeral, glad to see my mother, as I looked out at her, relax into my words as I spoke. My brothers and foster brother carried him to the hearse, and we all followed it from the church he'd attended for more than fifty years, past the schools all his children and many of his grandchildren had attended or were attending, and then past the neighbors' farms and past our farm, out to the cemetery, where he was buried near friends and neighbors.

His life seemed to me to be all of a piece. He'd made a decision early on and stuck to it—to this town and community and family and way of life. He was shy and averse to being around people. He'd never, for example, been in our town's grocery store. I believe the only time he entered the clothing store was to try on a suit jacket for my sister's wedding. Otherwise Mom bought his clothes and brought them home for him to wear, even his shoes. He was happiest in conversation with small children. My mother told me about how one day, before he was to have heart surgery the next morning, he'd stood at the window of their house and looked out at the scrubby trees, flat fields, and swampland. "Some people don't like this country," he said. "But I love it."

Once at home, I struggled to get back to my stride. The whole world seemed tilted. I realized that it would always be that way and I would have to adjust, as one finally does with a great change. I had the gift of my children, who were charging ahead, fully engaged. For them, and in honor of my dad, I had to keep up—and in that way I slowly re-joined the world.

***

A Crooked Solace

It meant nothing, I know that. No message
from the other side or on high, no
reaching out from my father, newly
dead, just the monarchs taking
a long rest in the plum trees we passed
heading to our south forty
to bale hay or dig potatoes.

I knew it then, all that last half
of August after he died, every time
I drove a team past the plums
and a ragged banner of butterflies unraveled
from the brittle fists of the bent trees
so my horses twisted their ears askance
at all those flimsy wings
skittering up and around every time, every
time we passed the ripening plums my dad
had liked so much, heading
to work at what he'd taught me, I knew
it was just the world happening

around us sometimes
as if it knows our need.

Chapter 16

Richard again suffered all the hay fever and allergic reactions and coughing and sneezing fits that we were used to having accompany the harvest of our grains. The bins of the granary filled with our rye, white wheat, red spring wheat, spelt, and buckwheat. I spent a lot of time with the fanning mill, cleaning bags of each. I couldn't get anything clean enough to sell to the co-ops that bought our potatoes, unfortunately. We made small, private sales to understanding friends who were willing to deal with the too-frequent kernels that were still in their husks. I kept grinding our flour and using it for our own baking. We fiddled with the fanning mill and researched better ways to harvest grain crops. And then the potato vines started to die back, and potato harvest began.

I bought some small, decorative buckets made with wood slats. I'd been growing garlic, shallots, and leeks in the garden for a couple of years and now had far more than we needed of any of those root crops. I filled the buckets with representative samples of them, plus a few of each of our potato varieties, and included a flyer about our farm. On my next trip to deliver potatoes to the city co-ops, I dropped the buckets off at a few restaurants I'd looked up or heard of from my more epicurean friends.

To my surprise, it worked. Now we sold potatoes, and often these other vegetables, to some of the better restaurants in St. Paul. The chefs sometimes came out to meet me when I delivered and told me how much they liked our produce.

It was a change from our local farmers market. We sold our potatoes and vegetables there, charging somewhat higher prices than other stands at which the vegetables were not organically grown, and often we'd have prospective buyers come up, look at our beautiful wares, and move on. Sometimes I'd hear a comment about how expensive our stuff was, and I always wanted to say, "Yes, it's cheaper with the poison on it." But the other farmers were so nice to us, and I knew they used as few chemicals, if any, as possible..

Mainly we tried to expand the scope of what was on offer at the market. I brought edamame, for example, which just looked like what it was—soybeans—and I even tried having steamed samples on hand. People were horrified at the idea of eating a soybean out of its pod. Kale was another mystery vegetable. We went home with every bunch of it that we'd brought. Leeks! I had beautiful leeks. I'd figured out a way to grow them with from six to eight inches of pure white stalk before it turned to the beautiful green leaves, which I trimmed into a triangle.

"What's a leek?" I'd hear one of a pair say to the other.

"Kind of like an onion, I think," the other would say. And they'd move on.

I was a terrible salesperson, too. I wanted to sit and read a book behind the stand and only get up to exchange produce for money. I wrote descriptions of each product we had on offer and placed it beside that product.

To my surprise, people would rather ask questions than read the answers that were right there in front of them. "How do you cook this?" they'd ask. And I'd have to say out loud exactly what I'd printed out for them to read. It was annoying—but as I learned, people don't go to the farmers market to read. They want to interact with the farmer. Luckily I had the much-more-approachable Marian, who didn't mind talking to people. I was available as a reference.

***

I continued to attend the Land-Use meetings, and Louie, our town chair, asked me to become a member. I enjoyed getting to know the other members of the little committee. I particularly liked Ed, a dairy farmer who had little patience for those who would use good land in a frivolous manner, which meant other than for crops or as habitat for wildlife. We had the same goal—preserving farmland while allowing for development—and we studied our zoning and soils maps and read about what other communities with a similar bent were doing.

It was informal and earnest at the same time. It was the right amount of involvement for me. My neighbor Kelly was on the town board, which seemed like a terrible responsibility. I sat in my car one evening, preparing to leave the town hall parking lot after a Land-Use committee meeting, and I thanked my lucky stars that I did not have to be on the town board.

***

The day came when Ron arrived with his horse trailer and truck. Katrina and Lucinda would leave us. Much as I liked Ron, much as I knew they would go to a good home, I felt sick about losing Katrina. "You'll like it in Tennessee," I told her as I led her up from the pasture. "It's warmer. And

I don't think you'll have to work as hard as you've had to here." I told that to myself as much as to Katrina as she ambled up to the gate behind me, expecting nothing more than being harnessed to dig some potatoes or mow hay or something—not to be sent away from her weanling daughter and pasture companions, never to see them again.

Richard and I didn't eat meat, in part because of the cruelty involved in raising animals to be killed. But how much less cruel is it, I had to ask myself, to raise animals to sell to others, who might or might not be kind to them? If you raise a meat animal, you can give it a very good life, end that life humanely in one quick action, and then make the best and most efficient use of that animal's carcass, so it will not have died in vain.

When you raise an animal like a horse, you give it as good a life as you can, you train it to become useful so its next owner will value it, and you HOPE it will have a good forever home with this person who is buying it. But once you have sold the animal, you have no control over its destiny. It could end up in a better place than you had the resources to provide, or it could be passed on to a less careful owner, to an owner in dire straits who cannot buy good hay or provide good pasture, or to an owner who works it too hard or ignores signs of the animal's pain.

There was also no denying the fact that the horses lived in family groups. Our herd was large enough, and lived in a large enough pasture, to allow them to mimic, in some ways, the behavior of a wild herd. We could see how a mare remained close to her foal, even after it had been weaned and returned to the herd—even after she'd had another foal. Sometimes mares took a shine to their grandfoals—our old Fjord mare, Krista, raced whinnying after the truck that bore her adopted grandcolt, Rim, away.

And now I was loading Katrina into the trailer to stand beside Lucinda. She followed me trustingly to the front, where I left her and slipped out a side door. Ron and Richard were deep in conversation. I stood with my hand through the little door, holding onto Katrina's leg. Touching her for as long as I could. "We'll take care of Patience," I promised.

"I know this is hard for you," Ron said when it was finally time to go.

"Just please take care of her," I said. "I mean, I know you will. But . . . she's such a great horse."

Ron hugged me and got into the truck and drove away. I never saw Katrina again.

***

That winter, one of our septuagenarian town board members flagged down Elwood, one of our town's employees hired mostly to maintain the roads. Elwood, in his sixties, agreeably stopped and opened the door of the grader cab to hear what Cal had to say. And what Cal had to say was, "I'll learn you how to plow," and he socked Elwood in the face.

Elwood pressed charges, and the supposedly quaint rural news item got splashed from our town into the larger papers in the twin cities and possibly beyond, for all I know.

"You have to run for town board," my neighbor Kelly told me. "Cal's a great guy, but his judgment is slipping."

I was horrified. But my mother always said that if you want something done, you usually have to do it yourself. She said a lot of things, but that was the one I remembered at that point, and I glumly agreed to run for town board.

The evening of the January meeting at which nominations would be taken for candidates to place on the April ballot was brutally cold. The snow drifts in the fields and ditches made the night a little brighter as Richard drove us to the town hall. We had to park along the road, as the small parking lot was full, and our boots crunched on the snow as we met and mingled with others heading toward the door.

Inside, the old schoolhouse smelled of the oil stove, which was burning hard to get some heat into the walls and floor of the antique building. In spite of its work, the room was cool enough that the people crowding into it kept their coats on, squeezing into the old desks and onto benches and chairs, and clumping up around the edges to lean against the wall. Carhartt is a brand of tough workwear, and plenty of guys were encased in worn versions of it—insulated bib overalls, long or short jackets, even hats. Women favored nylon jackets and coats puffed up with nylon batting. Scarves and mittens, hats and gloves. We were armed against the weather.

One thing I loved about our new community was how much it reminded me of the community in which I'd grown up. My parents had been newcomers there, too, as I was in this one. And they'd been welcomed, as Richard and I had been. Sometimes I thought it was as if I had grown up in a play, and now that play was being put on again, but in a different place and with different actors. Sometimes it seemed that familiar to me.

On this night, unwilling as I was to be entering into this new role in the town, I could not help but take pleasure in the proceedings. Once it was the appointed time, Louie, in his usual plaid-flannel shirt, stood up and called for attention. He led the assembled citizens through the process of what was to happen that evening—nominations would be taken for the various offices, we would vote for up to two people for each open seat on the board, and those names would appear on the spring ballot in April, when the people would again vote, this time choosing the person who would represent them on the town board.

The work of the meeting could have been done in fifteen minutes, but because of the joshing, the jokes, the wags commenting here and there, it took longer and the laughter warmed us. We voted by a show of hands to carry out the voting through paper ballots, which took longer to find, for one thing, and then to mark and count. People chosen to count ballots were approved by the roomful of their neighbors. We could see them attending to their tasks, counting and checking each other's work, and no one felt the need to question their results. For all the good-natured ribbing that went on, no one forgot the underlying reason for the meeting. We were carrying out democracy. In our small patch of the United States, we were taking it seriously and doing it right.

Doing it right that evening in January meant I was going to be on the ballot in April, opposing Cal. Richard drove home through the scathing cold, and I looked out the window at the moonlit fields of this place I was coming to love. We talked a little about how I would campaign. "You should talk to Cal and send out a letter with your information on one side, and his on the other," Richard suggested. "That'll save on postage and paper, and people won't get so much mail." I thought it was a good idea, but I knew I was too shy to go talk to Cal and suggest it.

"I'll send out postcards," I said. "I can get them printed downtown."

***

The phone rang in the night. I swam up from sleep—surely it was a wrong number. Too cold to get out of bed unless I had to, I waited for the answering machine to pick up. "Maureen and Rich, this is Kelly," I heard. I slipped out of bed and ran for the phone, hearing our neighbor continue, "Pick up the phone. Your horses are in our yard, and I'm afraid they'll get onto the road—"

"Kelly!" I said, having snatched up the receiver. "We'll be right there."

"The yard is full of these huge beasts," he said. "Kind of amazing."

By that time Richard was pulling on his pants in the bathroom, and I joined him, dressing quickly. Kelly's house was close to the road. Cady's Lane was lightly traveled, but someone might be coming home from working the late shift and plow into a 2,000-pound horse. It would be fatal to both horse and driver, and it was awful to contemplate.

As we were putting on our boots and coats in the mudroom, we saw headlights on the driveway. Another neighbor, Tom, coming to tell us he'd seen horses on the road. "I'll get halters and a bucket of oats and go up the farm road," Richard said. "You go with him and start tracking them from Kelly's."

Tom slipped behind the wheel. As he backed the pickup, I realized he was just a touch inebriated. "I was just topping up a bit after work at the bar with Jim," he said, putting the truck into drive after backing into the snow pile at the edge of the driveway. Then hitting the snow pile at the other edge, then backing and hitting the edge, then into drive and now finally making the turn to get us onto the cleared driveway and down the hill. Luckily we didn't have far to go. Out on Cady's Lane, which was covered with a light coat of snow that had fallen in the past few hours, we came to a section that was solid with hoofprints. And then they stopped, and I saw that they moved into the ditch and then into the field beside the road.

"Thanks, Tom!" I said. "I'll chase 'em down from here!" I slammed the door, waved again at Tom as he revved the pickup, and realized that it was pretty cold, once a person was out of the warm truck.

It would have been a more difficult walk across the snowy field had not the herd of horses crossed it first, trampling the snow down for me. They were heading back across our neighbor's field toward our farm, which was encouraging.

I considered the fence situation. The way the horses were headed, they would come up against the fence and gate we put up to discourage people from driving up the farm road on Miller's land, then crossing his field and going onto our place to party and drink and scatter beer cans—this had been a problem for a while. If I could veer them a bit to the north, they would cross onto our place through the line of box elders we tapped in springtime. It was unfenced there. But then they would cross the buckwheat field and come to a fence—I had to get ahead of them to open the gate.

I hurried as best I could—there they were, too far to the south. They were going to encounter the boundary fence. I saw them pause and, to my relief, veer to the north. They filtered through the box elders and milled about on the snow-covered buckwheat field, pawing through the snow to graze.

None of them seemed surprised to see me. I fed them morning and night, almost always in the half light or full darkness. They knew the sound of my swishing nylon snowsuit, and I talked to them calmly, though I was feeling rather urgent inside.

"Hey, Stella. Hello, Elaine. Good girl. Hello, sweet boy." I ran my hands over them as I passed among them on my way to the fence ahead. There was a gate there, just a single wire connected by a handle—I only had to unhook it and take the wire back and hang it on the post. Then they could pass through and, with luck, turn right at the farm road and go back down into the yard, and from there we could get them into the pasture.

I unhooked the handle, wrapped the wire back as I'd envisioned, and then, with a casual nonchalance I did not honestly feel, began to urge the horses toward the open gate.

Diana took the lead, and they began to follow. Unfortunately, being blind in one eye, she was heading for the section of fence BESIDE the gate. She'd hit it, jangle the wires, and they'd all spook and run off. I could see it clearly. Diana, more than any of the horses, was used to voice commands. She'd been in a six-horse hitch, I'd been driving her a good bit over the summer and fall, and there was only one way to get her to turn—I was too far back to reach her in time on foot. "GEE, Diana," I called with as much authority as I could work into my voice. "GEE." She obediently veered right, the herd followed, and they all flowed through the gate.

Now my challenge was getting them to turn north toward the house. As I crested the ridge, I saw a darkness against the snow and realized it was a man, with a boy next to him. Kelly! And that must be his son. Perfect!

"Just stay where you are! Don't move!" I hollered, and they stood like statues. The horses regarded them suspiciously and turned away, to the north. Bingo!

"Just stay in the yard, ladies!" I yelled to the horses as they moved down the farm road toward the house. "I'll get you situated with some hay!"

Kelly, his son Noah, and I walked down the hill after the herd. Kelly and Noah kept the herd lightly pressed into the corral as I slipped among them and opened the main gate, and they poured through as if happy to be back. I put out hay to keep them occupied till we'd fixed whatever part of the fence they'd gone through. And now Richard arrived, sweating and out of breath. He had dragged his halters, rope, and heavy bucket of feed up the hill, across country, running wherever possible. Steam practically erupted from his collar, he was so wet and warm. "Did you find them?" he gasped.

Now there was the opportunity to talk. Noah's dog had awakened him to see the faces of the horses, curious and huge, at his window. He'd run upstairs to tell his parents about the wild horses looking into their house. Kelly laughed at his dream, until Kelly himself looked out and saw the immense dark shapes milling about on the lawn.

After thanking our helpful neighbors, Richard and I set out across the pasture to find and fix the break in the fence. By the time we finished and got back to the house, it was time for him to shower and dress and leave for work. At least he would have a story to tell his friend Eric in the next cubicle.

***

Later that winter, Richard took the kids to California to see his grandmother. His parents came down from Oregon to join them, and I stayed home to feed the horses.

The house was empty. Because it was winter, I just had a few horses to train and some writing to do. I didn't have the pressure of yard, garden, and field work hanging over my head. I cleaned the house right away, enjoying the knowledge that my work would not be quickly overturned by exuberant children and unthinking husband. When I used a dish, I washed it, wiped it, and put it away. The clear counters were such a pleasure to me.

My days were organized around feeding the horses morning and night, but I could fill in otherwise as I wished. No one else's schedule figured into mine. Before falling asleep at night, I did a mental organizing of the next day's activities: feed the horses, go running, write, get a haircut, train so-and-so, feed the horses, write a letter to Inga, read, sleep. That is hardly the schedule to give a person goosebumps, but the uncluttered nature of it thrilled me. Just not having to make three nutritious meals each day, taking into account, to some degree, each person's likes and dislikes—this was heaven. I stopped at the grocery store after having my hair cut and picked up a box of Frosted Flakes, something I would normally never allow in the house. Cereal for lunch, three days in a row. Now that is living.

I knew my children were happy—I talked to them on the phone every day, and they were having a wonderful time with Grandma Shorty and their great-grandmother Ada. I knew I would see them soon. We could all make it through three days apart.

On the night before they came back, the sun went down and seemed to linger just under the horizon, burning red-orange. I sat on the window seat Richard had built for me against the west windows and watched the night slowly, as if it were engaged in a struggle, stamping out the fiery sunset. The sky was now black but for stars. The house was empty except for me, as it would be for—I calculated—about another sixteen hours. It would be good to have them back. For now, though, I was glad to read myself to sleep that night, not worrying at all about keeping Richard awake with the light.

The next day, I picked them up at the airport and expanded at the sight of their beloved faces. Joy and joy! But I couldn't forget how much I had liked being by myself. As I navigated the turn from one highway onto another, I told Richard a little bit about what I had done while they were gone. Nothing, of course, about the Frosted Flakes. That evidence had been carefully burned.

Not really thinking about how it might sound, I said, "I think maybe my natural state is to live alone."

In the gentlest, most careful, most tactful way, my dear husband said, reaching over to stroke my hair, "Well, we really like living with you." Which made me feel like the heel that I was, and I marveled in my heart that this good man loved me.

***

That spring, in addition to the mud, nights in the barn with foaling mares, a living room full of flats of onion, shallot, tomato, and leek starts, I had to campaign. In fact, one day when I was running on the road, Elwood slowed the big township dump truck he was driving and yelled down to me over the motor noise that I had better start campaigning. Of course he wanted Cal defeated—he hadn't liked being punched. "Don't worry!" I yelled up at him. I was worrying enough for both of us—but my concern was that I'd win.

I wrote up a short statement about myself, made it clear that I was running for the town board because I wanted to preserve farmland, and had it printed on a card I could mail to town residents.

Some I mailed. Some I delivered by hand, knocking on doors. Our town had a few rural subdivisions, and I found it most efficient to park the car and then walk, house to house, knocking, waiting, introducing myself, handing the person my postcard, and then moving on to the next place. Just as in teaching I had gathered myself to face the class each time I walked to the front of the room, that first rap on the door or press on the doorbell took some resolution on my part. Luckily for me, lots of people weren't home, for one thing. I could just drop off my card. In other cases, the person who came to the door was polite and even friendly. I quickly relaxed—until the next door presented itself.

The polls were open until seven in the evening, and then the paper ballots had to be counted by hand. Louie called me at about eleven that night, waking me to tell me that I was the new town supervisor II.

"Thanks, Louie," I said, sleep roughening my voice. "I think."

He laughed. "You'll like it! Serving the people! It's great!"

***

Our decision to sell more horses and reduce herd size was sometimes stymied by the appearance of Suffolk horses at auction. We could not let them be lost to the breed, as they might be if they were simply purchased by someone who didn't care about transferring ownership, breeding them to registered stallions, and then registering the foals. Now a team of registered Suffolk mares was going to be auctioned off at a sale a few hours away. Richard came home late that Saturday afternoon, and we were two horses richer.

Lily and Scarlett arrived a day or so later. As much as I didn't want more horses, these two mares were beautiful, quiet, and already trained to work. They fit easily into the herd. It was a good buy.

Because Krista was well into her twenties and we no longer used her for driving, riding, or breeding, we offered her to our friends Lynn and Jay. Their sweet daughters combed and petted and cared for her, and Krista enjoyed the attention.

One afternoon, however, Lynn called, distraught. Krista had fallen and was struggling to get up.

"It's tiring her out and she's not getting anywhere," Lynn said, worry apparent in her voice.

"Okay, I'll come now," I said, and I threw a shovel in the back of the car and drove over.

Lynn met me and showed me poor Krista, flat on her side, legs straight out and occasionally flailing for purchase. She'd somehow slipped and fallen, her body coming to rest in the trough between the wall of the barn and the ridge of ice caused by snow slipping off the metal roof and piling up, being doused with whatever moisture dripped from the eaves, freezing and thawing, matting to a hard, building-long hump of ice. I knelt beside her and rubbed under her ears, speaking softly, breathing in the odd, familiar scent of her, which resembled, I'd long thought, diesel fuel. The white around her eyes began to disappear. "We'll get you out, girl," I said. "You'll be okay."

Krista's back was against the barn, her legs angled up over the hump—when she struggled for purchase, she got only air. I could see that Lynn had been chipping at the ice, to no great effect. "It just makes her struggle more," Lynn explained. "I'm out of ideas."

I couldn't think of much myself. How horses manage to get themselves into these predicaments is a mystery to all horse owners, I think. We needed something to haul her backward. Behind her a few feet was the barn door, out of which she had come. Lynn and Jay kept that shoveled so the door would open. Once Krista was there, she could roll onto her belly and get her legs under herself to get up.

But how to haul her backward? I wondered what Richard would do. Picturing situations in which we'd had to move something big, I thought of a come-along. It's a cable that is wound onto a reel. There's a big hook on the free end. The reel is attached to a frame that has a hook on it. You unreel the cable and fasten the hook to what you want moved. You attach the other end to something strong and stable, such as a tree. Then you start ratcheting the cable back onto the reel, using a handle on the frame. It's a great tool for the small farmer. But I didn't have one with me, and Lynn and Jay didn't have one at all.

"Maybe Selmer has one?" Lynn's neighbor down the road, elderly Selmer, would certainly be at home, and it was worth a try.

Selmer Swenson's farm wasn't far away, but we drove down in case he wanted to come back with us. We knocked at the house and his wife, Muriel, her graying yellow hair piled on her head in a testament to the power of bobby pins, asked us in for coffee.

"Oh, thanks," Lynn said. "But we've got a horse down, and we're hoping Selmer can help. Do you know where he is?"

"Well, now, I don't know as I want two women going down to meet my husband in the barn," she teased, and we laughed.

"You know Selmer's safe with us," Lynn retorted. "He's crazy about you." Which was true. Selmer and Muriel had met late in life and love hit Selmer hard. "I think he thought he'd never find a woman who would love him AND the farm," Lynn said as we walked to the barn. "But he got lucky!"

Selmer's barn was from another time, another way of farming. Six stanchions—a time when a farmer could milk six cows and, with one enterprise plus another, this and that, make a living. There was Selmer, feeding his cows. He didn't have a come-along, but he offered to come along with us, and we laughed at his pun. He grabbed a rope from a nail on the wall, and he stumped along out to my car, using a fork as a crutch. "It grips better on the ice," he explained as we went. His accent was straight from my childhood among our Scandinavian neighbors. "Oh, now, are you sure you want me in the car? In these barn clothes?" He held back, his blue eyes worried.

"Look," I showed him, opening the door. "I have covers on the seats. You won't make a dent in the dirt that's been in this car."

He relaxed and maneuvered himself in, pulling the fork in last, angling it to fit. I slid the door closed and we drove back to Lynn's.

We wrapped a rug around Krista's hind legs and used the rope to pull—no luck. "How about using the tractor?" Lynn suggested. She went to get it—I was impressed. I still couldn't start up our Allis, and each time I had to drive it, Richard gave me a short refresher course as to which pedal was the brake, which was the clutch, and how to use the throttle.

I tied the rope to the tractor hitch and Lynn pulled ahead. Krista's hind end swung around and she thrashed herself halfway into the barn.

I had an idea. I'd just the other day read an article in one of our horse magazines—a guy's horse got into a swamp and was stuck. After all attempts failed, he'd pulled it out by the tail. I removed the rope from Krista's legs, wrapped the rug around the base of Krista's tail, and looped the rope around that.

Lynn looked at me in horror, but she pulled ahead. Now Krista was free of the doorway and the ice hump, but her legs were still uphill of her body.

Selmer and I looped ropes around her front and back legs and strained, flipping her over. Now she could get a purchase, get her feet under her, and . . . we held our breaths . . . stand up. Wobbling, but standing. "Oh, look," she seemed to say. "Some hay." She didn't even raise her head, just began nipping here and there at remnants of hay left on the ground from her morning feeding. As if nothing had happened, nothing at all.

Lynn drove the tractor back to its shed. As I coiled the rope, I remarked to Selmer that I was glad I'd read that bit about being able to pull a horse by the tail.

He cocked an improbably blue eye at me and said, in his lilting accent, "Ja, well, I guess they're fastened pretty good to that!"

***

I'd grown to love the town hall for its associations with the Land-Use committee. Our easy-going meetings were friendly, and we all shared the desire to preserve farmland. Now the old building had the same no-nonsense, creaky, almost-charm, but it wasn't my farmland-loving friends waiting for me inside on this raw spring evening. I'd left Richard and the kids almost as if embarking on a journey, longing to stay home with them. Our house, their dear faces, the quiet evening ahead for them—how precious it seemed.

Instead I gathered myself and slid out of the car and went into the building. I nodded—without looking at them—to the people lounging and talking near the doorway, and made my way to the front of the room. I sat down next to my neighbor and now fellow board member, Kelly, putting him between me and the audience. Louie sat perpendicular to us at the table, which was a big board from somewhere, propped on the old teacher's desk.

Our town board consisted of the town chair, Louie, who ran the meetings and did most of the talking, plus two supervisors, Kelly and me. The secretary, Janet, sat at the opposite end of the table, flanked by the town treasurer, Caroline. Secretary and treasurer could not vote.

Usually, town boards in rural areas consider such important issues as which grader to buy, how much salt-sand to stockpile for winter, or what carpenter to hire to do repairs on the town hall. Unanimous votes are the norm. In years past, the town board often met only to move to pay the bills, then adjourned the meeting and sat and talked about milk prices or the weather. Now, with the pressure of the twin cities moving out across the Minnesota-Wisconsin border, and the town's population increasing, our town board faced knottier issues.

Land was becoming more valuable as a place to build a house and less valuable as a field in which to grow crops. Corn prices were criminally low at two dollars a bushel, and no one saw that price ever increasing. The idea of preserving farmland for the future was, to some people, selfish and crazy.

Our town's zoning was designed in the 1970s by some prescient farmers and other residents. The state designated some soils as Exclusive Agriculture. Farmers who had these soils, and agreed to abide by the building restrictions on them, received one hundred percent of the credit offered to them on their taxes. One of the restrictions was that only a person who made a substantial part of his or her living through farming could build a dwelling on Exclusive Ag (EA) land.

Now an older farmer was selling off forty acres to a couple from the city. The forty acres was in the EA zoning. The city couple had no intention of farming.

The older farmer and his wife were also selling an 80-acre parcel to a man who planned to hay it, raise beef cattle, board horses, and work part-time as a farrier. He would not, however, be quitting his more-lucrative day job.

His request came to us first. The parcel he was buying was also in exclusive ag zoning. We had to consider carefully how we handled his permitting process because it would set a precedent for any that followed.

People were up at arms. Some wanted the board to hold to the ordinance. Others wanted the board to relax it. No one was ambivalent. At least, no one at that meeting that night was ambivalent. And everyone had eyes on the newbie, who happened to be me.

Louie called the meeting to order. Janet read the minutes of the previous meeting, Caroline read the treasurer's report. From there we took up agenda item after agenda item—none of great relevance—and with each one the tension in the room tightened.

The older farmer and his wife, Mr. and Mrs. Schnagel, were there in the way a couple of bristling porcupines might occupy space in a room. They were loaded for bear, as my dad would have said. Anything that smacked of government oppression of the little guy galled them—paying for the town's portion of the community fire truck? Putting in a new bridge on a road and causing the law-abiding residents on that road to have to make a long detour to get out to the highway? We at the makeshift table at the front of the room might as well have been the Gestapo.

Mr. Schnagel was stuffed into clean bib overalls and had a shiny red complexion that made me think of him as having at some point been boiled. His wife was brown-eyed and had the look of a woman who would be a lot of fun to be friends with, in different circumstances. Very different from these circumstances. Which were awful. The other couple to whom they'd sold land, the Svensks, sat near them, mirroring the Schnagels' fierce expressions.

Finally, Louie read the next agenda item: a request for Mr. Robert Cragness to build a home on Exclusive Ag zoned land. Now the room was silent, waiting for all the shoes to drop.

Mr. Cragness's lawyer, Mr. Hildseth, came forward. "Louie, my client has been patient so far—"

"Yeah," called Mr. Schnagel. "How long is a man supposed to wait?"

Mr. Hildseth continued, "Very patient. He'd like to get started on building his barn so he'll have a place to store his hay." The attorney carried on in this vein, dwelling heavily on the agricultural plans of Mr. Cragness, who sat quietly and watched the proceedings.

"I wonder if Mr. Cragness would consider, if he gets a permit, siting the house back further in the woods, rather than out on the crop land?" I asked.

Mr. Cragness leaned forward and answered my question politely. "It's too wet back there. I thought about that, but it just won't work." Mr. Schnagel erupted angrily. "There's no requirement in that ordinance about where to put the buildings!" His face gleamed pinkly around his little piggy eyes.

Mr. Svensk joined in, emboldened by the anger he felt around him. "Now you want to dictate where the house can go? How about making up some more rules while you're at it?"

I sat back, trying to place Kelly between me and the rest of the room. It didn't work, but just having him between me and the audience felt comforting. I was taking any port in this storm.

"The problem is," said Kelly, patiently, "we just don't have a precedent yet, and we don't want to set one that gets us to interpreting this ordinance in a way it wasn't intended."

"It wasn't intended to hurt the farmers!" Mr. Schnagel roared back at him.

"As a matter of fact," Kelly said—I loved him for his willingness to engage this guy— "you're right. It was intended to help farmers. And we want to interpret it in a way that honors that. That's why it's taking us so long to figure it out."

"You got that right," called a guy from the back of the room. "It sure is taking so long!"

Louie spoke up. "I wonder if we could allow it under a special use permit," he said.

"That way we can set some conservation measures as part of the conditions," agreed Kelly.

"Now there have to be conditions?" Mr. Svensk asked sarcastically.

"More hoops to jump?"

Kelly ignored him and continued to address Louie and me. "I feel as if he's met the spirit of the ordinance. Maybe a special use permit is the way to allow this. I move that we table it until next meeting so we can learn more before we make the final decision."

The room erupted. No one liked the idea, and everyone wanted us to know it. Nonetheless, I seconded Kelly's motion and we voted to pass it.

Mr. Svensk leapt from his chair and swaggered, knowing he had the power of the people in attendance behind him. (We had plenty of people who supported our cautious approach to interpreting this ordinance, but they were at home, trusting us to do our jobs . . . but it would have been nice to have them in the room, too.)

Just a few steps brought Mr. Svensk to the front of the room, where he thrust out his groin and demanded to be put on the next plan commission agenda. "Give me a number!" he shouted. "What the hell does 'substantial' mean?" Angry as he was, at Louie's polite request, he returned to his seat and sat down.

From there we went on to the rest of the agenda—to let the ambulance chief talk to us about the proposed rate hike for the service to our town, to discuss the recycling grant we'd received, to consider a dog complaint, to move to pay the bills. Finally, after midnight, we were done. The room had emptied out, and we sagged as we pushed our chairs back and stood up.

Janet, who had been quiet through the meeting, said, "That was the worst meeting I've ever attended. Sometimes I thought they were going to start using their fists!"

I was relieved to hear that I was not the only one. It would not have surprised me if they had charged the table.

"It isn't always this bad," Louie comforted me as we walked in the dark to our cars.

"Sometimes it's worse!" quipped Kelly. And then, "Not really. This was as bad as I've seen it. We'll get through."

That conversation was probably illegal. We were not allowed to speak to one another outside of the meetings. Because there were only three voting members of the board, any time two of us were together and talking about something related to the board, it was an un-posted, and therefore illegal, meeting. We could be fined for it. That thought hung in the air as we each bit back the rest of what we wanted to say and got into our vehicles.

Chapter 17

"I have an idea," Richard said, this time fully clothed and while we were outside walking up to look at the fences. "You know how the potato bugs are always worst in the middle of the field?"

"Yeah, I do know. I'm the one who pointed that out," I said. I remembered saying it. I'd followed that observation with a joke about how we should only plant potatoes at the edges of the fields. Ha, ha, as if.

"Well," Richard mused. "I was thinking about plowing strips in the field, just wide enough for eight rows of potatoes. Then leave a strip of grass, then plow another strip. All the way across the field."

Boing! I saw it immediately. "Oh, so no plant is more than four rows from the edge of the field!"

We theorized, after having attended workshops on organic potato production, that the beneficial insects lived in the vegetation at the edge of the field and moved into the potato field to prey on the potato-bug larvae. Making habitat for the beneficial insects throughout the field would give the plants a better way to resist the potato bugs.

"And then we can drive down the strips and spray the Bt if we want to, or a foliar feed, or both. I've been thinking about how to make a horse-drawn sprayer."

Well, it was worth a try. We hated picking off those potato bugs.

Richard's idea gave us another year of rotation, in a way. He plowed the strips into last year's clover/alfalfa/timothy hay mix, allowing the legumes on the unfarmed strips to fix nitrogen one more year. The next year, the hay strips were plowed for potatoes and the strips that had been in potatoes the previous year were planted to whatever was next in the rotation.

The sprayer Richard built was, to me, a marvel. He mounted a tank on an axle, made a place for me to sit, and put a pole on it for the horses. He used a solar panel from one of the fencers and mounted that to power a small pump that sucked liquid from the tank and forced it out into tubing that was tied to metal pipes that could be raised, for transport, and lowered, for use. He fitted more tubing to feed off the main tubes, so when the poles were lowered, four small plastic tubes dangled down, each one ending in a little nozzle. In theory, as I drove the horses down the grassy strip between the potato strips, poles lowered and the pump switched on, whatever we were spraying onto the potato plants would spray sideways out of the nozzles.

The first time we used it, we filled the tank with a mix of water and fish emulsion. We'd been told that the pores of plants are open widest right away in the morning, so we were up before dawn to catch and harness the team (my job) and assemble and ready the sprayer (Richard's). Richard did a trial run, switching on the pump, and a fish smell filled the air as a fine mist emitted from each nozzle. Perfect.

He raised the poles to the hooks he'd built in to hold them up, we hitched Stella and Jemima to the apparatus, and I started up to the field. Richard planned to follow with the pickup, in case repairs were needed and he'd have to go back to the shop for more parts.

I was driving the team past the hill paddock when we hit a bump, which joggled the long poles in their hooks and one of them slipped out, landing on the electric fence and bouncing a few times on the wire, which made a frightening sound to these horses who were well aware of what wire could feel like. I expected disaster, but both horses responded to my quick, "Easy! Easy! Good job, ladies!" and after the first startle, stood quietly as we waited for Richard to come and figure out how to get the metal pole off the electrified wire. Somehow the horses and I were not being shocked, as there were enough insulating barriers between us and the fence. Richard quickly remedied the situation by unhooking the gate, which stopped the charge from the section of fence upon which the pole lay, and he used some twine from the back of the truck to tie the poles into the hooks to prevent this situation from happening again.

Up in the field, I stopped the team at the end of a grassy strip, and Richard put the poles down. The nozzles hung perfectly between the rows of young plants. I switched on the pump and clicked to the team, and we moved down the strip. Everything worked perfectly. As the sun came up, the morning air filled with the smell of rotting fish. Later, down at the barn, the cats inspected me carefully after I'd let the horses back into the pasture. "There's something going on . . . smells so familiar . . . can't quite place it," they seemed to be thinking. I myself couldn't stand it and went in and had a quick shower.

***

Marian and I were walking up to pull scapes off the garlic plants. Scapes are the seed pods that garlic plants produce. Just as one breaks off the blossom end of a tulip after it has finished blooming so the energy goes to the bulb, which is where next year's tulip will come from, garlic growers cut or break the scapes off the garlic plants so their garlic bulbs will grow larger.

Ahead of us we saw an unfamiliar black shape in the field beside the farm road. "What is that?" asked Marian. "I don't know," I answered, wondering myself.

"Well," said Marian, in a tone that implied exasperation that we already had enough to do without adding anything else to our day's agenda, "I HOPE it's not a giant snail."

As it turned out, the dark shape was a piece of black tarp that had blown from somewhere, and we were spared the apparently time-consuming botheration of dealing with giant snails. At least for that day.

Up in the garlic field, we used scissors to snip off the scapes, saving some of them to make into pesto for ourselves and the rest of them to give to the local co-op to sell.

Our little co-op in town was a home away from home for the kids and me. We stopped there often for milk and whatever vegetables we didn't have in the garden or freezer, and I always had a plastic bag tucked somewhere in my purse for buying beans or rice in bulk. It was rare to go to the co-op and not fall into a conversation with friends, whether by a chance meeting or because one or the other of you was working there. It was a meeting place, a safe place to leave the kids if we had to run to do a few errands down the street, and a place to learn and talk about what was going on in politics, the organic business world, food in general, and of course just plain chit chat.

I was a cheese cutter—that was my volunteer gig for the co-op. If I could put in enough hours every month, we received a discount on our groceries. Cutting the cheese—many jokes were made about this job, believe me—entailed arriving at the store, checking the cheese cooler for out-of-date cheeses and ones that were nearly sold out, and making my list of what needed to be cut, wrapped, and stocked. From there I went on to washing the stainless-steel surface with the approved materials, getting the giant cutting board down from the wall and washing it, then hauling out the big wheels or blocks of cheese that had to be cut. I had to find the appropriate labels, the scale, and the super-sized plastic wrap. Once set up, I used a big, double-handled knife, holding it at both ends and using my weight to cut down through the cheese. I made the big cheese into little cheeses, and then I wrapped each small brick in two layers of the plastic, weighed it, priced it, labeled it, and stamped it with an expiration date.

Now that the kids could write, they were often pressed into service. Hands washed, closely supervised, they could actually be helpful. Marian wrote out the labels. I imagine customers wondered sometimes why the lettering was so uneven and labored on the labels, but she was careful to make her letters and numbers clear. Truman had a job he loved—using the stamping gun to stick the expiration-date stickers onto the cheese blocks. With their help, I could finish the week's job in under two hours.

We had to go into the store in the evening, just as it was closing, so we could finish in time to get home for the kids' bedtime. If we tried to accomplish the task earlier, while the store was open, we took up too much space and made it difficult for the employees to do their work.

Sometimes, to be honest, when I lay awake at night and thought about the crops, the weather, the pressures of development on our township, the worries I had about how I was spending my time and my life, I just got up and drove to town and used my key to open the store and cut cheese alone, with a light over me in the back of the store as I worked at, again, a job that was menial, helpful, mostly mindless, and time consuming, and that bore no relation to the ambitions I'd had for myself before farming.

***

One of the worries that kept me awake was our school. My children did not seem to be learning things they already would have had they been in public school. Marian was nine years old now, and the only impetus she had for learning the multiplication table was my proclamation that she could not get her ears pierced until she had her times table memorized up to the twelves. She was still spelling words as if she heard them in slow motion. She often wrote to her cousin Ashley, my sister's daughter, with something like, "Deeear Ashley, How are eeuuo? I am gist fine."

It was funny—in a way—but shocking. Ashley was a few years older than Marian, so of course her letters back were neatly written and almost always spelled correctly. But my friend Lynn's girls, who attended the public school, were writing as well, almost, as Ashley did! No spelling problems with them. They did not have the option of learning the times tables, either. Math class came as scheduled and they learned what was put before them. This Montessori school—was it too loosey-goosey?

On the other hand, the kids were well informed in other ways. Their teacher was reading the Odyssey to them, for example, and on drives home from school they would discuss the actions Odysseus had taken against his enemies, and the patience of Penelope, as if it were a current TV show.

They were confident in what they knew—and sometimes in what they didn't. One evening, when Richard was in England for a conference, we sat around the table after dinner, Marian knitting a scarf for her doll, Truman weaving a scarf for his stuffed bear, and me stitching the last bits on some mittens I'd sewn. At Marian's request we'd lit candles rather than turning on the lights as the night slipped into our house.

"Mom," Marian asked. "Does the word 'animal' mean only mammals, or does it also mean alligators and insects?"

"It includes insects and reptiles," I replied, surveying my work by the flickering light, inwardly wishing I could flip on some wattage and really SEE for a minute.

"I'm not a mammal," announced Truman, looking up from his cardboard loom—just yarn wrapped around and around as warp, and a popsicle-stick shuttle to pull the yarn through as weft.

"You are too," Marian said dismissively.

"No, I'm not," he argued, and both looked to me to referee.

"Truman," I said. "You're furry. You have warm blood—" he cut me off here, insisting he had COLD blood— "you have WARM blood. You are a mammal."

He bristled indignantly and retorted, "It's a free country!"

Well, we dropped that discussion. But you see my situation. I liked their goofy notions and wondered if their minds would remain as unfettered if they were in a conventional classroom.

***

A woman and her daughter moved into the apartment our family had lived in while our house was being built. Sherry, the mom, and Annie, the daughter, were quick to make friends with our family. Sherry, tall, thin, and angular, with a fused bone in her foot that made it difficult for her to take on jobs that involved much walking, such as waitressing, was interested in our farm. She found work as an assistant to elderly people who needed light housekeeping, and we were happy to have Annie join us for the times Sherry was at work. Annie was between the ages of Marian and Truman, and they were delighted to have a friend nearby who also enjoyed playing in the tadpole puddle out by the hill paddock or at the edge of the woods where they had a fort in a fallen tree.

It was fun to play at Annie's house, too, so there was a lot of back-and-forth between the two houses on our shared driveway. Between her house and ours there is a significant dip, where the narrow road passes between overhanging trees, and at the lowest spot there is a culvert under the road to let snow melt or heavy rainfall drain underfoot and then splash down the rocks on the steeper side.

Because of the trees and the damp, it is always a bit cooler in the dip. The kids decided it was spooky, and they called the area Ghostville. They decreed that one had to pass through with one's eyes shut. This resulted in a few missteps and near falls—it is a narrow road, after all.

The scariness of Ghostville was easily remedied by re-naming it Fairyville. A person could walk through Fairyville with eyes wide open, hoping to see one of the tiny sprites. Marian explained all this to me one day as we were driving to look at another horse.

She was too big for Spooner now. He'd been returned to his owners the previous autumn, and they'd found another home for him this summer. Peter, our vet, suggested one of his horses for Marian, now that she was bigger and a bit more experienced as a rider.

Country Mouse, or Mousie, as she was called, was an elderly Connemara pony—almost too big to call a pony, really, but old and slow enough to make me a bit less anxious about how much farther Marian would fall if she went off. Peter introduced her to Marian, who quickly fell in love with her. We went back the next day with a saddle and riding helmet, and Marian rode Mousie the couple of miles back to our house as I walked beside. Next morning, rather than running my usual loop on the roads, I ran to Peter's and drove the car home.

Mousie settled into her little paddock. We kept her separate from the herd because of her age and to make it easy for Marian to catch her for a quick ride—or to take her up to the barn and give her oats and stand on an overturned bucket and braid her mane while talking and singing to her as if she were a good friend.

In spite of Mousie's age and long career as a riding horse, I always worried when, after Marian brought her to me to do one last tightening of the cinch, Marian got on and rode Mousie up the farm road and disappeared over the ridge. I'd had enough horse-involved incidents and accidents in my life, enough casts and crutches and concussions and aches and bruises—I knew what could happen. For as long as she was gone, I kept an eye on the ridge, waiting to see the shine of her helmet, and then her head, and then Mousie's bobbing ears. I loved that she took these long rides and seemed to enjoy them so much. But I breathed more easily once she was back and telling me about the wonders of what she and Mousie had seen.

A new cat appeared—a raggedy stray who was able to make peace with Billy, Tom, and Mary. They allowed her to live in the barn, and after some gentle persuasion from Marian and Truman, she allowed herself to be petted and held. They named her Sarah, and I appreciated that she was so patient with being loved and hauled around in the wagon by the kids and their friends. I had things going on and couldn't pay a lot of attention to her, and then on top of everything else, my friend Laura had a sick kitten and was going on vacation. She asked me to keep her over Sunday and then take her to the vet on Monday morning.

The poor kitten was weak, and it was hard to get her to eat. I didn't want her to make the other cats sick, and Richard didn't want her in the house because of his allergies, so the kids happily made her a nest in the cat carrier Laura had left with us. The kitten stayed in there unless one of us was holding her. Sarah, the new cat, took an intense interest in the kitten and crouched by the wire lattice door of the carrier, gazing in at the baby. We were all sure she would hurt the kitten if given the chance, but Truman found, when he was taking the kitten out for some cuddling, that Sarah wanted to take care of the kitten. She licked her, purred, drew her close to her—we were charmed and loved Sarah more.

Early on Monday morning, though, when I got up to do my run, I found the kitten alone in its nest, dead. Sighing, I put her back down. I'd have to tell the kids when they woke up, and we'd probably have a funeral out in the area where we tended to bury our pets. I felt bad for them and bad for Sarah. She must have gone back to the barn.

An hour later I was back from running and was passing the kitten's nest when I saw that Sarah had been back. Draped over the wee kitten's furry body there was a warm, limp, dead mouse. As if Sarah had thought it might help. I stood over the little pile of death and thought about the old cat trying this one last thing to save the kitten.

Sometimes, as my mother said on the phone when I told her about it later, the world just breaks your heart.

Chapter 18

The Town Hall meeting at which the Svensks would expect to receive their permit to build a house on their land loomed like a crater in front of me. Board meetings took place twice each month. Every day brought me closer to that Monday evening. In the meantime, I studied our ordinances. I called officials in relevant departments at the county and state level and asked about what the terms in our ordinance would mean to them. I made a brief appointment to meet with our town attorney to ask him his opinion. I lay awake at night and fretted about how we could uphold the ordinance AND give the Svensks what they wanted.

I didn't much care for Mr. and Mrs. Svensk, to be honest. Mr. Svensk was overbearing, impatient, dismissive of our town's wishes to govern and regulate itself, and generally abrasive in manner. I knew less of Mrs. Svensk, though she had her husband's same opinions about our town. She'd made those clear in the comments I could hear—was meant to hear—from the back of the town hall.

But the Schnagels—I felt for them. They were dairy farmers, like my parents, and they'd worked hard for decades on that farm. Now farm prices were low, land prices were high, and they were ready to cut back. I'd heard they were planning to sell off their cows. They didn't need so much land, and it wasn't as if prospective farmers were lining up to buy it. Now they had a live buyer ready to put down good money on this piece of ground that they knew as no one else in the world knew, and the town board stood in their way. I could understand their frustration. At the same time, was it right for them to declare farming to be dead in our town because they wanted to quit? Shouldn't we all be looking ahead to the next generations as we made our decisions?

These thoughts chased each other around my head as I lay in bed or harnessed my teams or waited in the barn for a foal to be born. The morning of the meeting, I was thinking them again as I made the bed in our room, imagining how great it was going to be to climb into it that night, the meeting and conflict behind me.

"Four times eight is thirty-two!" Marian poked her head around the door and announced this mathematical fact to me, then scampered down the stairs. She was working toward getting her ears pierced. She'd checked on the price of having it done at the jeweler's—twenty dollars—and started saving. Just the past week, while putting away her clean underwear, I'd come upon her stash. Wrinkled paper dollar bills and a handful of change all stuffed into an envelope labeled "ear muney." Again, the worries about whether she should be at the Montessori school.

***

May evenings in Wisconsin are so beautiful. I walked from the car to the town hall steps, breathing in the long-awaited springtime air. Indoors, the springtime peace slid off my shoulders. There were the Schnagels, the Svensks, the also-angry contingent that had gathered around them, and the narrow path open to me to make my way to the front of the room, where I arranged my meeting materials—agenda, meeting minutes, budget tables—around me. From across the length of the table Janet smiled kindly at me, Caroline turned her head to say hello, and Louie and Kelly arrived close on my heels and settled into their spots. Louie gave all of us a quick, questioning look. "Are we all ready?" And then he called the meeting to order.

The item of new business was a public hearing for a couple purchasing land to build a house in which to live and in which they'd also have a photography business. Per the ordinance, they'd taken a petition around to the neighbors for signatures indicating no resistance to the idea. They'd brought it to our house and Richard and I had both signed. They were building on land zoned for houses, it is legal to have an in-home business in our town, and I saw no reason to oppose their plan.

Our neighbor Miller did, though. It was the first time I'd seen him, a handsome older man with an air that said he was used to being listened to. "Our house is across the road from this new place. The driveway will join Happy Valley Road just even with our living room windows. When cars leave his place, the headlights will cut right in!" He looked accusingly at us, as if we'd been planning on this all along.

Our heads all turned to Elwood, who had issued the permit for the driveway. He stirred in his chair and said, defending himself, "The driveway is just about the only place I could put it. It has to be set back from the bridge over the dry run on the one side and from Daryl's land on the other. And that hillside—the driveway has to wrap around and ease down to the road. I could look at it again, I guess, and maybe move it a foot or two one way or the other, but that's about it."

"That would be good, Elwood," Louie said. "It's no fun having a light in your face when you're relaxing at night."

"It's going to be car after car!" exclaimed Miller. "A foot or two one way or the other isn't going to make a bit of difference. He'll have people up and down all the time for that studio of his!"

"Are you really that busy?" I asked Mr. Hendrickson, the photographer. It was hard to imagine that kind of custom for a photography studio.

Mr. Hendrickson's hands were shaking. I saw him notice this and put them under the table. "N-nno," he said. "I really only have two or three shoots a day. And none of them at night."

"Now they've got those cameras that can shoot like a machine gun!" Miller insisted. "You can take a hundred pictures a minute. He'll be packing them in and we'll have no peace. No peace at all." Now Miller was sounding crazy, and opinions in the room full of people were shifting to the side of the Hendrickson couple.

"We have no objection to this plan," said a woman in the front row. "We are surrounded on three sides by the Hendrickson's parcel, and we are completely fine with having them build their house and David running his business out of it."

"As long as they realize I have a hundred head of cattle right next to them," our neighbor Daryl offered. "And they need to keep up their side of the fence."

Daryl's comment brought some rationality back to the situation. The mild-mannered Hendricksons indicated their willingness to keep up their fences and to have their proposed driveway re-evaluated, and I moved to grant them a permit. Kelly seconded, and we all voted to approve.

Next we considered whether Ruth Ahern should be permitted to give piano lessons in her home. In truth, most people just put out the word and gave piano lessons—but Ruth's husband Peter was on our plan commission, and he and Ruth wanted to be sure they were on the regulatory straight and narrow as long as he was making decisions for other families.

Mr. Miller again voiced an objection against in-home businesses in the town. Owen Frevert said he saw no correlation between what Miller had just said and the giving of piano lessons in the town. Fran Stark said the town needed more culture anyway. Ed Hanson said piano lessons had been given in the home from time immemorial. Once the comments from the floor dried up, Kelly noted that it had always been the philosophic position of the town to help farmers make ends meet by allowing in-home businesses such as piano lessons, and he moved to approve her application. Mrs. Ahern received her permit.

These were exchanges I loved. People spoke up from where they sat, crammed into some long-ago-child's wooden desk (in fact, Elwood's name was scratched into one of the desks, something he'd done as a boy attending school in this building), and everyone listened, evaluated that thought, and let it go or decided to comment or not. Louie maintained an open meeting at which the people felt free to speak up, and we at the front table listened. But now it was time to deal with the ugly part of the meeting.

All three of us on the board had made phone calls to various state agencies and studied the ordinance. We decided that our use of "substantial" could only be interpreted in light of the state's use of it, which meant on-farm earnings of $6,000 per year or $18,000 over three years. If Mr. Cragness could show us that he intended to meet that definition with his farming operations, then he did not need a special-use permit. He could receive a regular building permit.

Mr. Cragness had written out a business plan for the previous meeting. He presented it to us again, and we determined that he did indeed meet the state's definition of substantial. He was given a permit.

We could not sigh in relief, however, as next on the agenda were the Svensks—who were much more prickly and harder to bear.

Louie invited Mr. and Mrs. Svensk to the front of the room, and they seated themselves at the table. They looked sullen and hard. Again, I understood. They'd spent a lot of money on that land, and probably had a lot of plans for living on it, and now THIS. This dinky town hall, these dense officials holding them hostage over farmland preservation ordinances—it probably seemed like something out of The Andy Griffith Show, and all three of us were Barney Fife.

In fact, though, the three of us were doing our best to understand this ordinance and to interpret it in a way that would set a precedent for future decisions. We wanted to satisfy the Svensks—no one likes disappointing people—without caving in and opening the town up to further non-farm development that might hamper future efforts to farm the land.

But, at that moment, I think we all just wanted to be somewhere else. Louie called the public hearing to order and took up the Svensk's case.

Mrs. Svensk read a letter that said they loved the land and would do whatever it took to be able to build a house on it. They also planned to rent the fields out to Mr. Schnagel and to plant more trees. And to raise gingseng.

"Are you aware of how much it costs up front to get into ginseng?" an audience member asked.

"We've been reading about it," said Mr. Svensk. "There are different kinds of ginseng operations. We're going to a seminar on it next month."

Ed Hanson worried aloud that the proposed driveway took too much land out of production. "Can you move that driveway at all to save some of the crop land?" he asked.

Mr. Schnagel erupted from his chair—"It's none of your business where that driveway goes, Ed. I've been paying the taxes on that land for thirty-five years!"

Louie closed the public hearing and called the town meeting back into session. Kelly proposed tabling the Svensks' request until they had everything in writing, as Mr. Cragness had done. I saw the logic in that—having a uniform application process to the board, whether it was in the ordinances formally or not, would help both future applicants and the board as this situation came up again. As it certainly would.

In the now-raucous atmosphere of irked, annoyed, peevish residents, most of them taking their cues from Mr. Schnagel and Mr. Svensk, we moved and voted to table the request and moved on to the next item of business.

That meeting lasted till well after midnight also. We all sighed as we packed our papers and folders into the bags and cases in which we carried them to and from the town hall.

"At least we don't have to worry about turning off the stove," Janet remarked. It was warm enough now, this third Monday in May, to do without it.

***

Our spring work consumed me when I was not being anxious about town business. We'd got the wheat, spelt, and rye planted back in April. Richard found a cultipacker at an auction, and I thought it was the cat's pajamas. Imagine two stacks of checkers. Tip them onto their sides and run an axle down each cylinder. Each checker can rotate separately on its axle. The axles are connected so the two stacks are right next to each other. Rather than each checker being flat on the outside, the checkers are pointed. The stacks are offset. So when the cultipacker is pulled forward, the first stack rolls along, each cast-iron section (or checker) making a crease in the tilled soil. Then the next stack rolls along and the creases made by its sections are exactly off from the first creases.

Most seeds need a nice, fine, firm covering around them so they are in contact with soil moisture and their roots are not exposed. A cultipacker breaks up soil clods and packs the soil firmly around the seeds while also creating a light crust of soil that makes it difficult for weeds to germinate.

It took four horses to pull our ten-foot drill with the cultipacker hooked to roll behind it. If it was hot, or if the horses were not in working shape yet for the season, I'd plant with the drill and four horses, then switch to the cultipacker and do it separately. The finished fields looked like raked zen gardens with all those lines rolled into the soil.

We planted between five and six thousand pounds of potatoes in late May, early June. Lily and Scarlett were go-to horses for me, along with Margaret, Stella, Jemima, Belle, and Diana. I'd started Tug and Blue driving during the winter and hoped they would be the team of geldings I wanted to keep for a while. They were each still too green to be driven together, but because they shared the same type of squat, broad body, I figured they would be a perfect match.

"I want to plant buckwheat in that last field," Richard said to me as we were getting ready for bed one evening. "I got it plowed today, but will you disc it tomorrow?"

The next morning, after getting the kids to school, I harnessed Belle and Tug and put them to the disc. It wasn't a big field, and the day was pleasant. Belle was related to Katrina and Lucinda, but she was very different in build—low, squat, wide. She had a humble, willing personality, in contrast to the way they'd seemed to suck the attention (particularly Lucinda) from the herd to themselves. Tug worked beside his mother like a young hero. I was pleased and proud. He had a fun, happy way of going and was a deep brick-red color that darkened almost to black when he was sweaty. I imagined Blue would be much the same. They were going to be a well matched and wonderful team—my go-to team, especially during foaling season, when so many mares were either ready to foal or had just foaled. We liked to let a mare have a month with her new baby before putting her back to work.

I unharnessed them, put them back in the pasture, and before moving on to the next task, I wrote Tug's name on a strip of duct tape and proudly stuck it to the board that held up what was now going to be his dedicated harness. He was part of the working herd now, our Tug.

One night soon after that, we had a furious thunderstorm. I woke several times to the crashes and booms and flashes of light. In the morning the world was washed clean of the humidity that had been dogging us for a few days. I was up at the barn, feeling light and happy in the fresh morning, when I noticed two dark, still shapes in the pasture.

On the damp, closely cropped grass, Belle and Tug lay on their sides, noses almost touching. They'd probably been grazing nose to nose when the lightning hit the ground near them, stopping their hearts. Their hoofprints were deeper on the sides toward which they fell. They still had grass in their mouths. They were perfect, their shining June coats smooth as glossy river currents and then circling to whorls like eddies; the dark hooves; the lighter, longer hair at their fetlocks—as if some sculptor had studied two horses and made perfect replicas. They were so still.

I knelt and stoked their heads, tried to close their eyes, ran my hands over their perfect, perfect bodies, held their feet as if I planned to trim their hooves now that they were being so cooperative. It took a while to absorb the loss. I had to call Richard at work to tell him, and I eventually did, but first I spent some time in what seemed like a church to me that morning, what with the clear light, the clean, short grass, and the big, still, silent horses so recently alive.

Our neighbor Carl came over with his backhoe to bury them later that day. By then they seemed more dead—I didn't watch the process. I had to harness Stella and Jemima to mow the first hay of the season, and as I did so, I let my hands linger on their warm, living sides. I pulled that tape off the board with Tug's harness on it, and I crumpled it up and threw it away. Same with Belle's.

***

Truman turned seven in early May. His friend Sam came to spend the night, and we took them out for pizza. On the way home, Sam asked us what we planned to give Truman for his birthday. "Oh, I don't think he needs anything," Richard said. "He's got about all he can handle already."

This elicited the expected reaction from the boys. "No!" Sam exclaimed. "He wants a black mountain bike. He can't keep riding that old bike of Marian's!"

Knowing full well that we had already paid for exactly such a bike, and it was waiting for our trip to town the next day, I said, "Well, that's too bad. No black mountain bikes at our house."

Silence in the back seat. At home, they soberly got out of the car and Truman straddled his trusty blue hand-me-down (which had been through a number of children even before Marian had received it). "It just doesn't feel right," he said plaintively. Sam began to weep in solidarity. We should all have such friends, right?

Truman had a good birthday morning breakfast, and I made the cupcakes with which I planned to make the cake he'd requested—he wanted it to look like baby ducks swimming on a pond—and at just about the time the bike shop would be opening, Richard suggested that we take a little ride in the truck.

Oh, the joy when the bike shop owner, Adam, rolled that small, shining, black mountain bike across the floor toward Truman. Truman ramped up his ramp construction, building them ever bigger, longer, higher, allowing himself and his bike to shoot up the sloped side and then hang, however briefly, before landing the bike safely, one hoped, some feet beyond. This was just the beginning.

***

That summer was wet. We'd study the weather forecast, make the decision to mow hay, and then it would rain. Well, we'd say, not so bad to get rained on when it's still green. Then I'd ted it—run a machine over it to fluff it up and make it easier for the air to get around the grass, alfalfa, and clover—and possibly rake it . . . and then, of course, more rain. Rinse and repeat, ha ha ha, went the joke.

On the positive side, I could use the three-year-olds I was training to ted and rake hay. The work was not hard, and it was repetitive—around and around the fields we went. Before long they walked with their heads down and bobbing along, glad to stop and stand when I asked them to. Blue worked so well with Jemima, his mother, that I paired him with Elaine, who had been working well with Margaret, until Margaret foaled her beautiful Aldo and had a month of maternity leave. Agatha, Rosalie's foal, was coming along nicely, too. So the wet weather had at least one good thing going for it—lots of work for the young horses, and they all benefitted.

***

My friend Lynn was on the publicity committee for the co-op. "What about a farm tour?" she asked me one day. It wasn't something we would enjoy, but if it could help the co-op, I was willing.

"Not that many people," I said to Richard. "Probably just the co-op staff, mostly."

The day of the tour, we worked madly to make the place look nice. I'd mowed the lawn and around the garden the day before. I harnessed Lily and Scarlett and cultivated potatoes all morning, then finished the job with Elaine and Jemima that afternoon. Richard tidied around the barn. Cathy came out and worked with the kids to spruce up the garden. Then Marian had a realization.

People were coming TO OUR FARM. There was something she'd always wanted to do, which would never work because so few people ever came up our driveway, and that was a lemonade stand. The idea galvanized her. When she realized that Ellen and Jayne would be at our house for the whole event, her face flushed. It was almost more than she could absorb.

She and Truman divided the opportunities. She and Ellen would sell lemonade. Truman and Jayne would sell yogurt popsicles. Phone calls were made. "Ellen and Jayne are making signs," she told Truman excitedly.

She constructed a small table from a box covered with a towel and upended five-gallon seats for the vendors. She and Truman made the popsicles—I always had paper cups and a bag of wooden sticks ready for that purpose—and she readied supplies for the lemonade.

After that, she sat at her stand, looking dreamily into the distance. When we needed her, we knew where we'd find her.

I had Lily and Scarlett harnessed and hitched to a wagon when people began to arrive. We were shocked at the number of cars coming up the driveway. "I thought you said . . ." Richard trailed off. There wasn't much we could do about it now.

All evening I drove the team around our fields as Richard rode the wagon and explained our rotations and methods to the interested public. Down at the house, Lynn and Jay gave tours of the garden and orchard. Our respective children sat on buckets and raked in the cash, selling out of popsicles fairly quickly and then taking turns running to the house to make more lemonade.

By the time the last car had left, we were all tired and a little amazed. The kids were delighted, as you can imagine.

***

One rainy day in August I looked into the kids' rooms and was completely disgusted at the mess. "We are cleaning rooms today," I decreed. "Truman, you first. We'll do it together."

I gathered up some paper grocery bags—things to throw away, things to recycle, things to give to the second-hand store in town—and waded into the fray. Truman worked along beside me, his usual cheerful self. "What is this?" I asked about a plastic batting helmet suspended by string over a chair.

"See?" He sat jauntily in the chair, untied the string attached to a board lashed to his bed, and lowered the helmet onto his pertly held head. He maintained his upright, batting-helmet-holding pose for a moment, then pulled on the string, which went over a pulley he'd suspended from the ceiling, and raised the helmet. Then re-tied the string and bounced up from the chair.

Well, I did see. Such a useful contraption. I let it be.

In the meantime, Marian was busy in her room. Sometimes she visited us, stopping in the door to tell us breezily that she was really busy in her office. I could hear her in her room, talking to someone, and when I investigated, I found her sitting at her desk, looking out the window and talking on her cell phone, which was a Yosemite Sam Pez dispenser, extended as if to add more candy. "All right, that's good. I'll see you soon. Good-bye!" and she'd snap the dispenser together as if flipping a phone shut, and then make a frustrated face as now the Sylvester the Cat Pez dispenser rang and some other problem arose that she would have to solve.

Back in Truman's room, it was time to face what might be under the bed. I sent him to find out, and he agreeably slithered in and began tossing Lego pieces, nails, old Legomania magazines, string, socks, underwear, wrinkled pieces of paper, etc., out into the room at large. As he did so, he maintained a steady, cheerful chatter. All I could see of him was his left foot, one bright eye, and part of his face, which he kept turned toward me as he felt blindly to his right for more stuff to pitch out.

"Did you know, Mom, that T Rex can run faster than Bigfoot?" he asked. "Sammer says so. I think his dad told him. I might be in a bike race someday. Do you think we could get chickens again?"

It was hard enough to keep up with the junk he was tossing out, much less the thread of his commentary. I could only nod and agree, keeping my hands busy and thinking, remember this. Remember these funny kids.

***

The first day of school arrived. I got up early enough to run my loop and be back in time to awaken the kids. When I got back, however, I found them already up, washed, dressed, and waltzing clumsily together in joy at the idea of going back to Heartland. Truman had wet his hair and combed it to the side. Marian wore the dress I'd made her for the previous Christmas. For all my doubts about whether the Montessori method was right for my kids, I could not deny that they really liked school.

I would like to end this tale here, with my children dancing together and me standing, still not yet forty years old, in my house, laughing at them. What a good mother I seem, how strong a family we appear to be. Even the twin towers have time left in which to rise above New York City in a country that is not officially at war.

Chapter 19

What happened next in the story is that life carried on. We continued to raise and sell potatoes, garlic, leeks, and horses. I grew accustomed to feeling dread all day on the first and third Mondays of the month—town board meetings were held on those evenings, and there was always some kind of conflict that had to be dealt with. Richard worked and traveled and farmed. Marian rode Mousie through the following summer and showed her at the county fair. Truman souped up his bike ramps and bike. We began to receive bicycle parts catalogs in the mail, which he read from cover to cover, poring over each spread of cranks, chains, sprockets, and pedals.

Richard and I decided that the kids should attend public school, starting in the fall. Marian would go into fifth grade. I held Truman back; he could have entered third grade, but I felt his reading and math levels were more suited to starting the second grade. This decision was made easier because one of the best teachers in the district, Mrs. VerBurg, was able to take him into her multi-age class, where he'd have her for both second and third grades.

Marian was worried about starting at a new school, and I understood that very well. I'd never had to switch schools as a child and had always been grateful for that. Luckily, Annie, our neighbor girl from just the other side of Fairyville, had been attending Rocky Branch Elementary since her arrival in the district. "You just take it one step at a time," I said to my quietly worried daughter on the morning of the first day of school. "We'll all walk down to the mailbox together, you, me, Truman, and we'll meet Annie on the way. Then we'll wait for the bus. Then you'll get on and sit by Annie. When you get to school, follow Annie off the bus and into the school, and you know how to get to your classroom because we went there last week. Then sit down and soon class will start, and after that they'll just tell you what to do."

She nodded soberly, took the lunch I'd made for her, and we all began our walk down the hill to Annie's and onward. It all went as I'd said and when she came home that afternoon, she was relieved. "I have homework," she told me, "and I'm going to do it right now." I was relieved myself.

***

On the morning of September 11, 2001, Richard stayed home from work. We had to get rye planted as a cover crop on the big field. I went for my run, woke up Marian, and walked with her down to the bus. She was in middle school now and her bus came earlier than Truman's. Then the same routine with Truman. I caught four horses and harnessed them and had them tied in the barn as Richard organized the grain drill, the seed rye, and the cultipacker, then started washing potatoes as he waited for me.

I was in the house getting something when he came in, eyes wide. "One of the World Trade Center towers just fell down," he said.

I knew that couldn't happen—I'd stood on one of them back in the late-seventies one night, thrilled to see the carpet of lights stretching to the horizon, the vast blackness of the rivers and sea. Those buildings were built to last.

But Richard said he heard it on the radio. I turned on the TV. And sure enough, he was right. And as we watched in horror, the second building fell. We learned of the plane hitting the Pentagon, and the plane that flew into the ground in Pennsylvania. "All those people," I said, my voice quavering.

"They start work later on the east coast," Richard offered. "A lot of them weren't there yet."

I knew he was trying to make me—and himself—feel better, but how many times had he left early for work over the years we'd been together? It was far more often than not.

The news reports started to sound similar, and we realized that nothing was going to be solved in the next hour or two. We had horses harnessed and waiting; we had rye to plant. We turned off the TV and went out to the farm.

I drove the team around the field, drilling in the rye, and when I got back to the pickup, Richard filled the drill again. We had a radio that ran off a crank, and he'd crank it up and we'd listen as the horses rested, their panting making a backdrop to the horrible news of the day. How surreal it seemed, the carnage and vicious hatred that had led to it, and all around us the most beautiful September day, the sky clear even of jet contrails, the marsh hawks hunting over the hay and potato fields, our main herd of horses grazing far off in their pasture.

The next morning I had an errand in town, so I drove Truman to school. Everything seemed raw and oddly wrong, though for us nothing had changed.

"I don't understand," Truman said, breaking my heart as I saw him struggling to make the words, "I don't understand why they DID it. Those guys."

"I know," I said, struggling myself. "All I can think is there were nineteen bad guys. And there were hundreds of fire fighters who ran into the buildings to help the people. There are more good people than bad ones. Just remember that."

At the store, the cashier had red eyes, as I'm sure mine were red. We didn't say anything, but as she handed me my change, we both shook our heads. There really wasn't a need for words.

***

September 11

The first river fumbled along
following gravity down a crease
no one had seen yet because

nothing had eyes at that point in evolution,
which was still just a word
no one had thought of yet

because there were no brains, no eyes
so no one saw how the river
kept waiting for the rest of itself to catch up

but gravity is the big boss and does not
care if we love that blue china dish
that was used at Great-grandmother's wedding,

too bad if it slips from wet hands while being
washed, gravity values it as much as anything else
and wants it, whole or in parts,

as it took those who jumped or fell
from the towers, the law is the law to gravity—
just ask the rivers,

who keep leaving, and leaving,
not even at rest in the oceans' gyred currents,
the same salt as our tears.

***

Marian's birthday party that January was remarkable to me. She invited several of the girls from her class at school. Always before she'd had her classmates, mostly boys, from Heartland, and she'd received books, educational toys, and art supplies. Now, for her eleventh birthday, I can only characterize her gifts as being pink. And glittery. Nothing educational, all of it flimsy. She received each silly geegaw with joy. The girls' pink, slippery jackets were piled in our mud room like a heap of melting party mints. They exclaimed and laughed, said "like" more than I'd ever heard it said before, and ate as if their parents had not fed them for a week.

I think it was how much they laughed and ate that made me feel as if these girls were going to be a lot stronger than their present predilection for pink ruffles and glitter indicated. (And I was right.) Richard and Truman made their cautious way back into the house after the guests departed. Marian stowed her pink nail polish and various necklaces and earrings and bows and ornate hair accoutrements beside her rock collection, animal-identification cards, and the plastic, partially dissected human torso with its removable organs on the shelves in her room. When I went in to kiss her goodnight, she was still glowingly happy with the events of the day.

***

That spring we welcomed more babies out in the barn. Marian and I snuggled down to wait for Lily to have her foal one cold night in March. At about 2:30 in the morning we watched Eleanor's slick, dark form, wrapped in the stretchy sac, emerge from Lily's backside and come to life on the straw. Lily turned her beautiful head to see what was happening, saw her ridiculously scrawny, wet, bewildered filly, still attached by the umbilical cord, and whickered softly, that low sound sending a charge through the air, a kind of a thrum that established an understanding—Lily loved this little beast and would do anything for her.

The next morning little Eleanor looked fit. Around noon we let Lily out into the paddock behind the barn. I scooted Eleanor before me, and Richard held Lily's halter to soothe her as we brought the two from the dark barn into daylight, where the brown grass and partially melted snow made a rather inhospitable setting for a foal's first experience of the world outside. Eleanor stayed at Lily's side; we put down some hay for her, and I spread straw as a bed. One of the most endearing features of a foal, in my opinion, is its utter confidence that it is THE MOST IMPORTANT being on earth. The mares do little to dissuade their babies from this notion—just like we humans, they know it won't be long till the foals find out otherwise all by themselves. In the meantime, if a mare is eating hay off the ground, her foal will almost certainly plop down onto the nice, soft hay, leaving the mare to nibble at what protrudes around the sleeping baby.

Lily was looking thin, so I wanted to make sure she got enough hay. As I worked, the rest of the herd noticed Lily. They were just on the other side of the fence from her, most of them mares or youngsters, and they lined up along the fence to watch with gentle curiosity this new arrival. I could imagine them saying, "Oh, Lily, she's so cute! How did it go? Hard time? Oh, I'm glad you're well. What a sweetie!" There were about twenty of them ranged along the fence, each one with the same charmed expression, though I suppose I'd be hard put to explain how a horse can look charmed.

***

One early evening as I was preparing for the town meeting that would be held later, Lynn called to tell me that Krista was dying. Lynn and her family had moved her onto another farm, where Krista would have closer attention—Jack and Julie's kids were grown up, and though Jack still worked in the city, Julie was home all day and spent most of her time outside with her sheep—and Krista.

But now Krista was lying down and could not get up, and that is never a good thing in a large animal. I called Janet to let her know I'd miss the plan commission meeting and probably be late for the town meeting, and I drove out to Jack and Julie's.

Krista lay flat on her side. Her eyes were bright. She sometimes flailed to get up. She was so thin—her backbone was like a handlebar, and at the times when she'd struggle to rise, I grabbed it and tried to use it to pull her onto her belly.

Surrounding her were Lynn, her little girls, Jack, Julie, their adult daughter Kiki, and me. Shortly after I drove up, the vet arrived. She carried a large syringe full of pink liquid—all of us knew what that was for.

She listened to Krista's heart, took her pulse, and looked at her gums and eyes. "I don't see a real reason for this other than her age," she said. "There isn't anything I can recommend to help her out. But this has to be your decision."

Well, technically Krista still belonged to me. It was my call. I nodded. "I think she's had a good life. Let's not prolong any suffering."

Jack, one of the world's kindest men, wept openly, making big, racking sobs. Lynn's girls ran into her arms. Kiki and her mom each clung to Jack, crying, as the vet administered the drug and Krista simply died.

I stroked her one last time and thanked everyone for the kindness they'd shown this old horse. Jack assured me that they'd take care of her body. Then

I had to get to the town hall for that night's meeting.

Louie, our town chairman, called it to order. As usual, I wanted only for it to be over. I was chilled from sitting for an hour in the cold, and from hugging a damp horse. I was sad. But the procedure of gaveling the meeting to order, calling for the secretary's report, treasurer's report—the ritual calmed me.

Leaving the town hall that night, turning off the fuel to the stove, flipping off the lights, Louie asked me, "Would your kids like to bottle-feed an orphan lamb in case I have a ewe that rejects hers?"

Well, I knew the answer to that one.

Within days Louie pulled into our yard in his blue pickup and got out, carrying an impossibly small box. "Had a ewe reject twins this morning," he said. There were two in that wee box?

Marian and Truman led him to the barn, where they'd made a pen from straw bales, and scattered straw over the floor, making sure they were ready for their new lamb when it came. They hadn't expected to get two! Louie put the box down on a bale and opened a flap. The frowsy head of a lamb emerged, followed by another. The two sweet babies elicited "aaaah" sounds from Marian, Truman, and me. Louie pulled out a bottle, and one of the lambs latched on. Louie handed that lamb to Marian. Truman stepped forward and soon held the second lamb as it sucked enthusiastically on the nipple of its bottle. The kids sat on a bale, lambs on their laps, bottles tipped up. The faces of my children were enchanted.

"I'll be back in a few weeks to get them," Louie explained to them. "They just need some help getting started, and then I'll bring them back to my pasture and they'll be part of the flock. You can visit them."

After that, Marian and Truman fed Hector and Sox first thing in the morning, before heading down the driveway to the school bus. I did the late-morning feed and then picked up the kids at the end of the driveway to get them home a bit sooner so the lambs could get their afternoon milk. They were fed again just before bedtime.

As they grew, we could increase the amount of time between feedings. The trips to the barn with full bottles were beginning to lose their luster, I saw, and I had an idea. I drove T-posts into the lawn and stretched a net fence from the house to the picket fence on the west side of the yard. This made a small paddock for the lambs, and the kids constructed a little straw barn for them. Now feeding the lambs was a matter of simply opening the west windows of the living room and holding the bottles out. The lambs, much larger now, stood and sucked like a couple of woolen houses on fire.

One day Marian was tired and stretched out on the window seat to read. From the kitchen I could see the woolly faces, curious, at the glass just inches from her shoulder.

"Why don't you feed me?" I asked. Marian looked in surprise at me, then saw where I was looking.

"You sillies," she said affectionately, and she opened the window a crack to scratch behind their ears.

***

Our neighbor Kelly had helped me get started keeping bees in the first year or so after we moved onto the farm. As I had in years past, come late July, I borrowed his extractor and clamped it to the picnic table. My bees had made enough honey to share with us, and it was time to get it into jars. I showed Marian how to use the heated knife to cut away the wax cappings over the honey cells. Truman helped me arrange the frames in the extractor, four at a time. Then both kids had to sit on the picnic table to steady it as I cranked the extractor, getting it up to a speed that would fling the honey against the walls of the extractor barrel. The picnic table rocked, in spite of the kids' combined weight—plus mine—and I cranked with one arm and held onto the extractor with the other, trying to keep it from rocking itself across the table, much as a washing machine in the spin cycle can leave the wall and "walk" across the laundry room.

Then I slowed the cranking and allowed the frames to stop spinning. They had to be reversed, so the honey on the other side could be flung out— again with the cranking and rocking and steadying. The air filled with the scent of honey. The kids stuck out their tongues, as if they might taste it. We worked in a cloud of bees—they'd found us. When I opened the spigot to let the honey run out into the fruit jar I positioned beneath, an amber bar descended and appeared to melt and fill the jar. The table filled with jars of honey. Marian busied herself with saving the bees from drowning in it— some had become entrapped and were floating near the surface. She used a spoon to extract them, settling each bee onto a clean place, then gently squirting it with water from a bike bottle. Eventually most of her little squad of bees was able to fly away, much to her satisfaction.

Later I'd bring in the jars, carefully wiped of their stickiness—but always, nonetheless, sticky in places all the same, no matter how careful I was—and put lids on them and stored them in a cupboard. Over the months ahead, when I opened the cupboard, that honey caught whatever light there was and glowed a little bit, as I imagine the gold in Tut's tomb did when they made that hole in the wall at first discovery.

The dense plenty of my storage cupboards gave me satisfactions that are akin, I suppose, to those an investor might have in reviewing quarterly reports on successful ventures. Five-gallon buckets of wheat berries and other grains, waiting to be ground into flour, quart jars of tomato sauce and honey and maple syrup, pint jars of jelly and jam made with plums or berries or crab-apples picked on our farm—over the years it became just how I did things, but in those first seasons of harvest, I took a lot of pleasure in the results.

***

Swarm

I saw it in the elm above the bee yard
like a lady's softside handbag slung
from a branch, the weight of itself pulling
it into a shape as if it carried, oh, maybe
a wallet with coins for the parking meter
or some apples to eat
on the train. Ten
thousand bees of one optimistic mind, let's
go and they lift off, cloud, congeal
on a branch, thicken it, drip
from themselves, wait
in a satchel-shaped pulse
for the scouts to tell
them where they will go.

I studied the tree, imagined
myself climbing a ladder, saw
in hand to cut the branch, get
that swarm, re-hive it. Our longest
extension ladder, set in the bed
of the pickup truck, maybe that
might reach to where I could stretch
the saw up and at that point
I thought let it go, seeing myself
not as I once was, ready to climb
ladders freely, but as I am now
having heard from scouts
how far and hard
is the journey and where
it will end.

***

"Where's Marian?" I asked Truman one day after school that fall. He'd trudged up the driveway to where I often met them at the top of the hill, beyond Fairyville.

"Her and Annie decided to stay on the bus," he said. "It goes around the loop, so they'll get off next time."

I was annoyed, because I didn't know what time she would get home, so I worried fruitlessly until I saw her heading up the road at five, when I was out feeding the horses.

When I came in, she was waiting for me in the mudroom. I asked, very somberly, "Marian, why are you so late in coming home?" She clearly knew she was on thin ice, and she hated to be in any kind of trouble, so she very seriously and earnestly tried to explain.

"We had a new bus driver, and Annie and me decided to stay on the bus and show him the way, but then, then we remembered that—we don't know the way."

I imagine this bus driver thinking, oh, good, these kids can tell me where to go. And then their earnest admittance that they never ride the bus this far and didn't know which way to turn. Just great, he must have thought.

I was able to keep a straight face and make my point—that she should remember that people worry about her when she isn't where she should be, when she should be there, and to remember that in future. Later I told Richard what she'd said, and we sagged with laughter. We are not long on brains in our family, but we do like to help.

***

Richard felt that we needed a lean-to on the north side of the barn. I was on board—storing our equipment inside would make it last longer and break down less often. I was less thrilled about the work ahead, but as it turned out, the kids were now big enough to be really useful.

By now Richard had an auger to dig the post holes for the long poles that had to be set. It ran off the power take-off on the tractor and did not require the attention of two people holding it upright as it drilled down. I helped get the poles in and held them upright, level pressed against the wood, monitoring the bubble to make sure each pole was straight as Richard filled in the holes, shoveling and pounding dirt, trying not to bump me as he worked.

We didn't have a good way to get the trusses up. Richard solved the problem with ladders and me. The ladders were uneven, one taller than the other. We had to make do with that fact. We positioned the truss on a board, which was balanced on two piles of bales. Then the two of us got under the board, rested the board on our bent shoulders, and lifted, thus lifting the truss. Then we had to climb our respective ladder, taking care not to be faster or slower than the other, bending our bodies to keep the board level as the different heights of the ladders came into play. The truss was lifted over, then rested upon, a board Richard had screwed to the side of the barn.

The weight off my shoulders . . . what a lovely relief, and to have not dropped it or fallen or tipped it sideways so that it hit Richard on the side of the head—but then we climbed back down the ladder, caught our breaths for a while, and did it again. Over and over. Move the bale piles twelve feet west, re-set the board, position the truss, get under the board, do the lift—oh, forgot to move the ladders! Ha ha. Carefully bend ourselves back down to rest the board and truss back on the hay, move the ladders, start again. Fun times.

The kids could not help with this. In fact, we were so worried about something bad happening with them around that we did the rafter work while they were at school. Richard went in late or came home early from work. But once we had the poles set and the rafters secure, the kids became Richard's earnest and loyal helpers. I went back to my own work—harvesting the garden, cleaning garlic, canning tomatoes, studying for board meetings, writing, etc. Marian and Truman hammered enthusiastically at the siding, saving Richard hours of boring time he'd have had to spend doing that himself. Or, worse (to my mind), I'd have had to do while he cut the siding to size, built doors, and accomplished the more challenging work.

The magazine I'd worked for had expanded. I wished often that I could have been part of the discussions as they started another magazine—this one for very young children, and then another, this one with a science bent, also for younger children, and then a magazine for older children. It was exciting. I was happy for the company and for the staff. But I was a little bit glum, I have to say, about not being part of it.

Until one day I received a phone call from the editor of the science magazine for younger children. She asked if I would write the parent companion to each issue. It would be published with the magazine as an insert, and I'd offer suggestions and tips for parents to enrich their children's experience of the articles in the magazine.

Well, I certainly would do that, Debby! A job that would pay money and that did not involve sweat or dirt or sitting in the town hall? I found the work to be absorbing and interesting, and I enjoyed doing it much more than I liked nailing siding onto a barn frame. So as Richard and the kids worked outside on the barn, I studied and wrote. We were all happy.

Richard was becoming increasingly unhappy, however, at work. He was concerned about a chemical his company was making that he felt could have a bad effect on the health of animals and possibly people. He wanted to be home, farming. He was frustrated at being across the ocean or several states away as problems cropped up on the farm, such as broken fences, escaped animals, an outbreak of potato beetles, or a big, unexpected order for potatoes that meant the kids and I had to be out picking up and washing spuds late into the night.

We talked a lot about whether we could make it on our farming income. We'd lose health insurance—that was huge. His pension, once he was old enough to begin receiving it, would be much less than if he finished out his career at his job.

I was more concerned about the financial aspect of his decision than he was, it appeared. What I dreaded was the stress ahead. As a child on the dairy farm, I'd seen how hard it was for my parents to make a living, no matter how hard they worked, when the weather didn't cooperate. A rainy June, for example, could cause enormous problems because we depended on hay for the cattle. We stored baled hay in our wooden barn loft, and it had to be good and dry for two reasons—so it wouldn't catch fire some winter night, having been put up damp enough to compost itself into flames, and so it wouldn't mold and be less nutritious. We needed three crops of hay, starting in June, and it all had to be dry. My dad's temper was not soothed by the vagaries of weather, I can tell you that. Now, sometimes experiencing similar stresses myself, I understood him better.

I didn't want that kind of stress in my life—it was one of the reasons I'd resisted farming. There were a lot of things I enjoyed about our farm, and one of them was that we could survive a crop failure because we had the income from Richard's job. We'd always worked as if that was not a consideration, but it really was. Now we'd forego that.

Richard and I went over our books from previous years and projected income for the next. He longed to be at home, rather than at the office or flying around the world to meetings. He wrote a stern letter to his company to explain his disappointment with their lack of response to his warnings on the chemical in question, and he resigned.

That first morning he was home, I really wondered if we'd survive the proximity. I was trying to make lunches for the kids to take to school and he was making breakfast for himself. He learned not to get in the way during those times, and we became accustomed to each other's routines. It was wonderful to have him home, in fact, and not traveling for weeks at a time. It was a decision we have never regretted.

Chapter 20

"There's a Suffolk gelding for sale," Richard said, lowering the paper he was reading. "I guess I should call."

Well, we did try to make sure no Suffolk horses were lost to the breed, so I agreed—sighing somewhat because I didn't want more horses on the farm.

Richard was smiling when he came to find me after he'd talked to the horse's owner. "It's Martin," he said. "He's been gelded. I guess he's kind of a handful, but probably nothing you can't handle."

And so it was that Earl and I struck out for a farm south of Menomonie where a Mennonite farmer met us and accepted my check and led us to the barn, where a familiarly big-rumped draft horse was tied in a milking stall. It was Martin—that back end and his blaze were unmistakable. I led him from the barn and, once he saw the sunlit door ahead, he bounced forward to the end of the lead rope—which fortunately was a long one—hauled me outside with him, and joyously bucked and plunged.

I laughed—he was so clearly happy to be outside. After a fashion I reeled him in and loaded him, and Earl drove us back to Baldur Farm.

Richard and I had talked about what to do with him upon arrival. Putting him in with the herd would result, we thought, in a lot of turmoil that could result in an injury to Martin or one of the other horses. Best to put him in a paddock within sight of the herd, but with a couple of fences between him and them. So, when we arrived, I led him to the hill paddock and showed him the situation with the fences, and then unclipped his lead rope.

He raced through the fences, whinnying as the wires snapped and fenceposts popped out of the ground or bent sideways. With a trail of hissing wire, he raced to the herd, and they—horrified at this new horse and the snaking wires he was wearing like a necklace—raced away. And he galloped after them.

Martin VanHuisen. All these years later, I still shake my head. I got him free of the wire, eventually, and just let him sort it out with the horses, some of whom seemed to remember him. I was pounding fenceposts back into the ground when Richard came home and surveyed the damage. There wasn't much to do other than fix the fence and hope we hadn't made a mistake in bringing this big, undaunted gelding back to the farm.

***

Truman's ramps for his bike had progressed from little slopes of plywood propped on two-by-fours to him balancing on his bike on top of the picnic table, then powering down on his pedals and shooting himself and his bike forward and down onto the grass. He asked if he could try BMX racing—there was an indoor track about fifty miles away.

In this sport, the racers ride small bikes. They balance against a gate, which snaps forward and down at the sound of the start buzzer. The racers burst forward and race each other over a series of obstacles carved and pressed into the track. The track snakes around sharply, and more often than not, it seems to me, the race is not won by the fastest rider but by the one who can stay out of the crashes and just finish the event.

Truman was good at it. He brought home trophies every time he raced. Now, rather than pore over the Lego magazines, longing to win "The Lego of Your Dreams" contest every month, he received bicycle-parts catalogs. He sat at the kitchen table and studied each spread. Frames, and all the different configurations a frame can take. Then cranks. Then pedals. The tiniest differences were not lost on him.

"Mom, I need clipless pedals." I didn't know what they were. He burned to have them, though. Through a combination of birthday money, a sale at the bike shop, and some assistance from his mother—grateful for his help with various jobs I needed to do—he was able to secure the pedals and shoes he wanted. Sure enough, he won his races the next Sunday afternoon and was convinced that now he needed a better chain.

"When will this end?" I asked, the question bubbling from my past. My parents gave me a bike when I was in second grade and I rode it, unquestioningly and uncomplainingly, until I was in high school and bought the neighbor girl's old ten speed, which I then rode for years. Performance, for me, was just being able to get from point A to point B.

Truman didn't understand the question. End? It would never end. Every month, he was learning from the parts magazines, there were improvements to every single piece of a bicycle. He wanted them all.

***

My friend Inga was working on a project—she was going to adopt a baby girl from China. Over a couple of years I followed along through emails and phone calls—the initial consultation, visits from the case worker, talks with women who'd successfully adopted a child, the excruciating wait once she'd jumped the final hurdle. And then the call, the blurry photograph of a wee girl seated in a highchair, and the amazing, astonishing knowledge that within weeks she would be at home in her room in Inga's house in America.

"Will you come with me to get her?" Inga asked. "I'm not sure I know how to handle a baby on my own."

Well, that was an easy question to answer.

I'd never considered that one day I'd have an opportunity to visit China. And of course I was eager to meet Inga's daughter. Flying to China, though—that was imposing. Being so far from my family for nearly two weeks! And in August, the month in which finally my garden was at its peak and I was harvesting and canning and freezing the fruits of my labors. The potato fields were beginning to die back; we needed to get started digging spuds. As I fretted, Richard said, "Just go. Everything here will wait."

The kids were excited for me. China! I was touched that they did not question Inga's desire to have me with her as she learned to care for a baby. Of course she would! To them, I was an expert in that department. Myself, ten years out from having Truman, I wondered if I remembered anything at all about what babies like and need.

As it turned out, Nicola needed very little once she was in her mother's arms. She was given to Inga in a hotel room, and we were allowed to take her back to our room. Inga held her in front of her as we walked, stumbling in excitement as we noticed the dimples in Nicola's perfect cheeks.

Nothing says I-am-your-mom like feeding the child, so Inga sat down with Nicola while I prepared a bottle, something I had never done for my own children (because of being lazy and a skinflint more than because of lofty ideals regarding breastfeeding). I dumped the formula into the bottle, only to be shocked when it slid out the other end. There was no bottom to the bottle, and Inga had to take her attention away from Nicola long enough to tell me where to find the little plastic bags that would hold the formula, once I had it mixed.

Inga needed little assistance in learning to care for Nicola. I showed her how I gave baths to babies. I dunked Nicola's little hand into a cold beer after she'd stuck it in Inga's hot soup and scalded it a bit. Otherwise, my main use was in carrying and holding things as we made the treks we had to make from one government office to another to obtain the passports and documents necessary to allow Nicola to leave China with us and for her to become, officially, Inga's daughter.

Which, in fact, had happened—unofficially but truly—in the first hours they were together. "I'd walk through fire for her," Inga told me with a kind of awed seriousness after spending the first night alone with her daughter. Love just kicks down the door.

This happened again, only more quickly, two years later, when Inga adopted Marianna. I tagged along this time as well, but my babycraft was not needed. Inga was an expert by now, and Marianna and she bonded in the first moments of their meeting. I was again useful as a porter—babies require so much stuff!—and as a babysitter, so Inga could get out and use the workout room in the hotel sometimes. Marianna looked small, fragile, and meek. But this was a ruse. Whereas Nicola, at nine months, had very little strength in her legs—so little that I put a pillow on my lap to elevate her to being able to look out the window at the boats on the Pearl River when I had her to myself—Marianna at seven months had springs in her knees. She bounced on my lap, but her greatest enjoyment was in being thrown as high as I dared and then caught again. I did this next to the bed, so if I fumbled she would have a soft landing. She loved it so much that I began throwing her up with a twist, so she'd do a twirl in the air before falling back down. Over and over. When Inga came back I showed her the trick, and by her skeptical expression I could see that she was not really into it. Luckily, I discovered that another of Marianna's loves was looking at herself in the mirror over my shoulder. She could do it for an hour at a time and would probably have enjoyed a longer session, but I was tired of staring at the wallpaper in the hallway as she enjoyed that cute baby in the mirror opposite the wallpaper.

I missed my family intensely both times I was so far away. There was no way to communicate other than through email, which could only be accessed through the hotel's business office, and it was expensive. Seeing Richard at the airport to pick me up after the first trip reduced me briefly to tears. I was tired from the long trip, I missed Nicola, and at the same time I was so glad to be back. How beautiful my children and house and farm looked after that time away.

***

Martin VanHuisen adapted to his position in the herd. He was allowed to be among them, but he had to take direction from Margaret and the other mares. He turned out to be a terrific worker in harness. He was such a fast walker that I hitched him with Margaret, who was the only horse we had who could keep pace with him—she was taller by a hand.

Winter came, and after a snowstorm Richard said he couldn't plow us out because he couldn't get the tractor started. I was impatient, so I harnessed Martin and hitched him to the wooden V-plow Richard had made some years before. It had a narrow setting, which I used first to open up a path down the driveway. Coming back, I opened it another notch, widening the path. Going back down, I opened it a bit more. I had to sit on the cross piece to weight the plow or it would ride over the snow rather than push it aside. Martin hit an icy patch and began slipping, dancing a bit to keep on his feet. Another horse might have panicked, but he just glanced back over his blinder (most of our harness bridles had saggy blinders because the horses rubbed their faces on each other while they rested) and his cheerful look said, "Don't worry. I got this."

When we needed a single horse to pull the wagon with the straw chopper on it through the garlic field, mulching the garlic, I harnessed Martin. He hated the slow pace, having to stand around until Richard—back on the wagon—gave the go-ahead to take five steps. I tied the lines back to the wagon and just stood by his head and talked while we waited, and this seemed to calm him—though the look he gave me after we walked together, but stopped after only five steps, was one of resigned bafflement.

In winter we stored as much of our equipment as would fit in our barn, crowding it in a complicated jumble with a path here and there so we could reach the potato storeroom and the ladder up to the hayloft.

One night Martin got out and found a jackpot in the barn—I'd left some oats in a bucket in the harnessing area. He licked that clean and ventured further into the barn, ending up in a corner. Richard came out in the morning and was greeted with Martin's delighted whinny and then watched in helpless dismay as Martin made his eager, clumsy way, tripping and stumbling over the equipment, breaking and bending metal as he hurried to get to Richard, his good friend.

He got out at will, no matter what we did. And then, in the morning, he'd see us and charge over as if he had thought we were lost forever. His huge rump and his feet like tin buckets, his eyes open so wide that his forehead had a permanent wrinkle—I wanted to murder him sometimes and yet often wished I had two of him.

One afternoon I was fixing a fence Martin had broken the night before. Truman came home from school and joined me as I tried to drive a post into the frozen ground.

"Mom," he said, as usual waiting for me to answer before he continued.

"Mmm-hmmm," I said, lifting the post driver to slip over the T-post and positioning it, then pulling it downward—CHUNK CHUNK CHUNK. The post slipped sideways on the ice and I swore inwardly.

Truman winced, understanding my frustration. Gosh, what a good kid he was! But he wanted to tell me something, so I waited before trying again.

"I can bound," he said with some little bit of pride. "Wanna see?"

"Oh, I sure do," I said. Who wouldn't?

He then proceeded to run around the pasture, leaping—well, bounding— over frozen piles of horse manure, finally fetching up beside me, puffing and proud.

"That's some good bounding," I said. "Yep." And I started pounding the fencepost again, this time with better success.

At dinner that night, as usual, Truman sat in his place across the table from me. I noticed his flushed cheeks and asked, "Truman, you look like you might have a fever, you're so rosy. Do you feel warm?"

He appeared to give it some thought and then said, matter of factly, "Well, I have been bounding."

***

Truman was such a good kid. Marian was beginning to be sour as she grew closer to becoming a teenager. She complained about having to weed in the garden or mow the lawn or pick potato bugs. Truman, I sometimes thought, was like the best dog you ever heard of—just whistle him up and he was there, ready to help, cheerful, interested, and useful. On a hot afternoon when I was digging up garlic alone, because I hadn't thought it would be right to ask them to come out in such heat after being helpful in the garden all morning, Truman found me and helped bundle and carry the bulbs and their long greens to the garden cart, which we piled high before hauling it, straining together, up to the barn where I hung them to dry.

"Truman is always . . . chosen," Marian complained bitterly one time when I was talking to her about helping out more around the house and farm.

"I don't choose him, he volunteers!" I said. "You could do that, too!" I could understand, though. So much of what he did was mechanical—helping to hitch some machine to the tractor, running to get the air compressor tank for his dad, holding extra tools as Richard or I worked on whatever implement we needed to use next. Marian was just not interested in that part of the farm.

Marian's superpower was that when push came to shove and we needed to get something done before a frost or a rain or an event, she dug in and worked beside us, complaining at first and then settling into the job and, I could see, taking satisfaction in being part of the effort.

One autumn I had the garlic planted in a timely manner, but there hadn't been an opportunity to mulch it. Richard was away on a trip—he was still working on the chemical that worried him so much, and he often traveled to meet with other scientists or to meetings to learn more and talk about it. There was frost in the forecast, then rain, then snow. If I was going to get that garlic mulched, it had to be on that day.

It was a Saturday. I filled the truck bed with straw bales, piling them as high as I dared. I drove to the south forty and made piles along the garlic strip, returning to the barn for more as needed. Then I woke up the kids and explained about our day's planned activity.

"It's cold," I said. "Dress for it."

I walked back out to the field, leaving the loaded truck for them to drive when they'd finished dressing and having a bit of breakfast. By the time they arrived, I was through the first pile of bales. I showed them what I wanted them to do: Move the bale onto the strip ahead of where the mulching ended, cut the twine (they each had a knife), use a fork to lift the bale sections, and then shake them onto the buried garlic cloves. Repeat. And repeat.

It began to rain—a cold, wind-driven rain that spotted my glasses and found every opportunity to seep into and under my clothes. We worked on for a few hours, until I couldn't stand the misery in their pinched, cold faces, and I sent them back to the house. They hadn't complained even once. "I'll be done soon," I said. "Probably another hour or two. Thanks for helping me. You can take the truck back down."

"You're welcome," Truman said, unable to hide the relief in his face as he turned toward the pickup. "Okay, Mom," Marian said, also delighted at this turn of events. I watched the truck bump and roll its way back across the field and then out of sight, back down to the house. Then I turned back to my task.

I was good and cold by the time I finished. The rain had turned to sleet some time ago. Cold as I was, I took pleasure in the puffy-looking layer of straw over the long strips I'd plowed into the field with Martin and Damon. The garlic could probably withstand the cold of winter, but this blanket allowed it a bit longer to develop roots that would give it a head start in the spring. And the mulch would keep the soil from freezing and thawing and freezing again, which moved the soil around and could stretch and even break the roots of the garlic.

I rested my fork on my shoulder and bent into the north wind and its buddy, Mr. Sleet. It was a long walk back to the house.

Once there, I was surprised to be met by Marian, who searched my face to see how I was feeling and then showed me that she and Truman had set the table. She had a salad made for me, and a skillet warm on the stove with eggs in a bowl, cracked and ready to be scrambled and cooked for me.

I had a quick, hot shower and then enjoyed a hot meal cooked for me by my surprisingly competent and caring daughter. I was looking forward to a good crop of garlic, and I was beginning to see a glimmer of what might lie ahead as pertained to my efforts at parenting.

Or not. I wrote a letter to the editor about an issue in a neighboring township. They'd passed a one-year moratorium on factory farms, and the previous week's paper had featured several angry letters from people who didn't think that was right. My letter tried to make the point that small governments, such as town boards, need time to figure things out before making big changes to the landscape or the code book. But those who profess to love family farms, rather than factory farms, need to support the former through their buying habits. It wasn't unusual for me to write a letter to the editor, but it was unusual for me to receive much feedback on my published thoughts. For some reason, I heard from a lot of people, as I completed errands around town, and they were quite positive in their praise.

Moms and farmers are not much used to that, I guess, so I reported happily on it at the dinner table that night. "Sue, the cashier at the grocery store, said she always likes my letters because they're so thoughtful," I said, arching my brows at my children to show them how special their mother was.

Marian looked up from her spaghetti. "You're always thoughtful," she said, accusingly. "You always talk about your thoughts."

It had been bothering her lately—I talked too much about issues. Not enough about . . . well, I didn't know what else to talk about.

So I asked her, "What should I talk about? Lipstick?"

We all laughed. And then we did talk about lipstick for a while, oddly enough. A twelve-year-old girl finds that much more interesting than zoning ordinances. Even I know that.

Chapter 21

The kids were lucky we drove the pickup when we went to visit my sister and her family. Their uncle Bob had a new treadmill, and the enormous box was in the garage, waiting to be broken down and recycled. But, of course, Truman saw it and showed Marian, and they had to have it. We tied it down in the bed of the truck for the ride home. A few days later I was coming down from a potato-washing session up at the barn and saw that the kids were busy cutting holes into it and taping things into it and so forth.

"We're making an apartment house for the cats!" they told me. "But the floors keep collapsing."

"And then the cats run away," Marian noted. Well, cats are not stupid, I thought. But I left them to their work and went back to mine. When next I saw it, they'd stopped using tape and started using wire to hold up the various "floors" in the building. Billy had the entire second floor to himself but was looking rather ginger about it, and I couldn't blame him.

We were now selling to all the co-ops between River Falls and downtown Minneapolis. Then Whole Foods moved into St. Paul, and we got that account, too. It seemed as if all we did was dig, pick up, wash, and sort potatoes. One evening we looked at the forecast and were startled to see that the temperature the next night was predicted to drop to seventeen degrees. This would not only freeze any potatoes left on the surface of the field. It would also freeze, or at least damage severely, even many of the potatoes that had not been dug yet and were covered by dirt.

We'd dug potatoes that day and had left out a couple of rows' worth to pick up the following morning. That worked well for us, as Richard could get up and get started picking up potatoes as I got the kids off to school, walking with them down to catch the bus, and then hurrying back up to catch the team to harness and drive up to the field where we then hitched the four horses to the potato digger. That particular morning, Marian had a fever and stayed home in bed as I went through the routine with Truman.

We dug as many rows as we dared, and I trotted the horses going back down the hill so I could get them back into the pasture as fast as possible and myself back to the field to help Richard. I stopped to check on Marian, who was feeling somewhat better. Then, zoom zoom back up the hill to fill the crates with potatoes.

We could see Cathy walking toward us from across the field, and she'd probably never been so gratefully welcomed in her life. She settled into the job and we slowly cleared row after row of potatoes, filling the hayrack with crates and burlap bags full of spuds.

I met Truman at the top of the hill as he walked up from the mailbox after school, and I explained how he'd need to come up and help as soon as he'd changed his clothes and had a snack. I checked on Marian again—she still had a fever, but she was sitting up and reading, and I thought, well, if she can do that, she can clean garlic.

So as Truman got ready to come up and help us, I brought bunches of garlic down from the hayloft for Marian to clean for market the next day. I helped her set up a cleaning station in the garage, somewhat warmer than the barn, and left her with a radio and earnest thanks. We had an order for fifty pounds, plus an order for a hundred pounds of leeks—which I'd have to harvest and clean after we'd picked up all the potatoes. By then it was sleeting. However uncomfortable that was for us to work in, we took comfort in the idea that the temperature was unlikely to go as low as was predicted. A clear, dry night would get much colder than a cloudy, moist one.

I had a town committee meeting that evening but did not finish in time to make it to the town hall. Well after dark we all stumbled into the house, dirty and tired. We'd harvested 4,000 pounds of potatoes, plus the leeks. And Marian had the garlic order ready to go. I'd like to have been able to order a pizza for dinner, but none of the places in town delivered as far out as our farm. We had our usual—microwaved potatoes and salad. Somehow we just never got tired of eating spuds.

***

That winter I had six young horses to train. I decided to start with Aldo, who seemed to be the steadiest of the bunch. I wasn't wrong—he caught on quickly. So quickly and so well, in fact, that once I had Jed ready to hitch with a workmate, I chose Aldo as the breaking horse.

Marian had a school project that entailed observing a particular location in nature every day for a month. She chose a small pond on the way to the south forty, and soon it became a habit for me to have whatever team I was working with ready to take her up the hill once she'd come home from school and was dressed to go out into the cold. She'd tell me about her day as I drove the horses up the hill, and then jump off when I stopped them at the path to her observation area. I'd stop there again after a loop around the south forty and she'd come along shortly and ride back down with me. It was a pleasant way to learn about her day—now that she was in middle school she was less willing to share with us. Riding behind a couple of big red horses seemed to bring her home to the farm in a more gentle way.

It snowed so much that month, I could no longer just take the young horses out right away. Even if it had not snowed, the wind blew snow over the track the horses and sled had made the day before. I had to harness Margaret and Martin, who were taller and more powerful, to go out and break trail first. Then the younger horses could more easily pull the sled along,

Patience was proving to be much more difficult to train than the other horses. She was hard to catch, halter, lead, tie, handle, and harness. The other horses were amiable and willing. I tended to work with them first and shorted her on my attention. She puzzled me—when I came to the pasture, she stood facing me, her large eyes fixed on my face, looking as if she wished I would come to her. If I did try to pet her, the best she would allow was the flat of my hand pressed onto her forehead. I was not allowed to touch any other part of her. Putting a halter on her was nearly impossible because she would not let me move to her side. She was larger than any of the other horses her age and could do real damage if she wanted to. Whereas the other horses quieted down once they knew they were haltered, she became agitated. Because I had so many other horses to train, I left her till last and often just stopped for the day without working with her.

In time I did get her trained well enough to work in the fields, but it was never easy to harness her. She pulled back and broke the snap, the lead rope, or the halter with ease if she was tied up and something startled her. I took to leaving her untied and having someone stand behind her and off to the side, where she could see him or her and feel pressed, I guess you'd call it, to stay up where I could harness her. Each time I harnessed her was like the first—she had to be shown the collar, the harness, and the bridle, and allowed to snort her opinion of it before I could put it on her or sling it over her back. Once harnessed, she was fairly calm, and she worked well in a team. Getting her into the barn and harnessed, though, was such a chore that I tended to prefer whatever horse was next to her, rather than bring her up to work.

It wasn't till years later that I had a thought about what might be the situation with her. I was planning the spring vaccinations and figured I could jab all the horses right there in the pasture—simply halter them one at a time and administer the shot, then move on. But, of course, that would not work for Patience. She never stood for shots. "If I harness her, though . . ." I pondered. She always calmed down once harnessed.

It occurred to me that maybe Patience was on the autism spectrum.

Some kids are bothered by a tag in their shirt or underwear, or they have a problem with looking someone in the eye. They want companionship or even to be touched, but on their own terms. And that was Patience. It didn't make her any easier to handle, knowing that, but I felt less frustration in working with her.

Aldo, on the other hand, became my right-hand man. He and Jed were beautifully matched in stride and size. I called them my team of professionals, and Richard agreed with me that it was time I had a steady team that I did not have to think about selling. When I went to catch them and bring them in, Aldo would see me coming and walk away, as if he'd just thought of something he had to take care of over there. I'd carefully not look at him, but approach him almost accidentally, quietly singing, for some reason, Bridge Over Troubled Water. Almost as if it were unplanned, we'd meet and greet each other and I'd put his halter on, lead him over to Jed, halter him, and then walk back to the barn, the two horses ambling obligingly behind me.

One time, Sarah, Richard's older daughter, was visiting us. We were thrilled to have her there, and she was eager to help. I was busy with something and, after describing where she'd find him and how she'd know him (he has a small star and Jed doesn't, Martin has a blaze, Sam is the smallest one), she went up the hill to get Aldo. My head was bent over some machine or other when she came back and brought him over to show me and make sure she'd picked the right horse.

I looked up and had to laugh. Aldo was looking at me with enormous significance, a kind of suppressed glee in his eyes, and he seemed to tilt his head to indicate that he was with someone ELSE, not me, and he felt special! Maybe you think a horse can't convey those ideas with a look, and maybe you're right. I only report what I saw, and that horse was tickled to have the attention of our Sarah. I don't blame him—we all loved her so much and still do.

In the evening every few days Marian or I would go out to fill the stock tank in the geldings' paddock. There were from four to six horses in there, and in summer they drank a lot of water. Seeing someone down there by the tank with a hose, they'd pick their way down through the grass in that way horses have, as if they are swimming. What a project they'd make of slurping up water, as if this new stuff was really special.

The most fun part of the job was hosing Jed. They all enjoyed being hosed—we'd put a thumb over the opening and make the water jet out, and we'd give them a spa treatment with it. Eventually the horses would tire of the water and the hosing and they'd go back to whatever they'd been doing. Except for Jed, who could not get enough of being massaged with cold water from the hose. He moved to make the spray hit different parts of his body—here, now here—and no matter how often I told him he was a complete and total goof, he just soberly turned to get some hose action on his rear end. It was impossible to do that job and come away from it unsmiling or feeling stressed.

Horses and water—watching our herd play in the farm pond in their main pasture made me think it would be the best thing in the world to have hospitals and nursing homes built around bodies of water where animals come to have their drinks and cool off. The joy and the drama! By now Damon was our herd stallion and led the parade up to the water in the pond paddock. He'd lower himself into it as if he were an old guy taking a schvitz, groaning his pleasure, and then rise up, shining with the water and pulling himself on to the shore to watch the rest of the herd enjoy the pond each in their own way. Some got in and pawed, pawed, pawed at the water, making an incredible noise as they wetted their bellies. Others got in and lay down, no nonsense about it. The foals were fun to watch—they'd teeter at the edge and then hop in, wade out till it was over their knees, and then, wide-eyed with the excitement of it, lower themselves till the water went over their backs. Then—wow! They'd pop up and scramble back onto the banks, thrilled with their bravery and this new sensation of feeling cool on a hot summer evening.

***

Reflections

In New York City there were lots of windows,
the kind that reflect passersby and I saw myself
uncomfortably often,
as I do not here on the farm
where I see myself mainly in the responses
of animals—birds, say, flitting back into bushes, waiting
for me to fill the feeders and happiest to see me
walk away with the empty scoop. Or the cat, who
climbs into the loft where I am cleaning garlic,
taking the ladder one rung, one rung, as
I do, the aluminum ringing lightly under his paws,
So I'm looking over there when his head clears the loft floor
and his yellow eyes do a critical scan
of my situation, which again to him looks
like another of my time-wasting schemes. Or deer,
who freeze in place to gape

at my clumsy run, who leave the garden
with a kind of eye-rolling sheesh when I scold
them out of the kale. Or turkeys, whose pink heads blush
blue as they scatter and fade into the woods.
Or the horses, who see me coming and wait calmly,
their brown eyes soft in their interested faces, all of them
beautiful enough that when I walk among them
I feel beautiful too, something easily proven untrue
by big shiny windows but we choose, don't we,
our mirrors and I'll take these around me that say
I'm nothing special in such entertaining ways.

Chapter 22

My mother signed up to go on a tour sponsored by the local bank one autumn. She'd been born and had grown up in Canada, and though she'd earned her US citizenship decades earlier, she still loved the country of her birth and had visited all except the maritime provinces, which were the subject of the bank's tour.

She returned from the tour, happy to have seen the beauty of Canada's eastern seacoast. But she told me her neck had hurt her so much, she knew that was her last trip. "I won't be doing that again," she said, and I wish, looking back, that I'd been less flip in my response. I can still hear the sadness in her voice. She always loved to travel.

She'd been tireless and strong—mother of seven children plus a foster son, active in the community, a registered nurse at the local hospital and then the clinic. I was home from college for spring break one time and watched her step over a baby gate, barely breaking stride, with a husky grandson tucked under one arm and a wad of sheets for the washing machine in the other.

All that changed when she was struck, months later, with rheumatoid arthritis. She became, in her own words, an old lady. The pain must have been beyond belief—her hands turned to eagle talons, nodes formed on her elbows, knees, and feet, she could no longer hold a job or even keep the house tidy. She told me that when the minister was visiting her one day, all she could do was watch a mouse dart across the kitchen floor. "I couldn't even set a trap," she said.

Her strength hadn't just been physical, and her inner spirit kept her going over the next couple of decades. She was again able to knit and crochet, after a point, and she turned out hats and blankets for the grandchildren that kept coming along. She was an active committee person in the church. She could once again cook and do light household chores. She cared for my dad as he declined and died.

And now my mother was in a nursing home. The twenty years of steroid drugs to manage her disease had taken an awful toll. Her bones were dissolving. Her spine was having trouble holding her up. The brace devised to help with that cracked her pelvis. She was in the town nursing home, where she'd been an RN at the time she'd been struck down by the disease that led to her being there.

After a few months she was well enough to leave the nursing home but not well enough to go home and live on her own. My sister and I discussed it at length, the two of us, a hundred miles apart, pacing around our homes with cordless phones tucked between lifted shoulder and ear, doing dishes, sweeping the floor, folding laundry, trying to get our mom-jobs done before we headed off to our respective work as nurse and farmer.

"She needs someone to come and live with her, or she needs to live with one of us. And you have the only house that's on one level," Andrea pointed out. I agreed with that. Mom wanted to go home, to HER home. We wanted that for her, too, but we both had families who needed us and work that we had to be on premises to do, and we just couldn't—or felt that we couldn't—take ourselves away for a month or so as she gained strength to be able to go home and be on her own.

I looked forward to having her with us. My kids loved her; she'd never spent an overnight at our house. Their cousins who lived within a mile or two of her (and my dad, when he was alive) had the pleasure of her coming to their concerts, wrestling matches, and basketball games. She was more involved in their lives because she lived nearer to them, not because she loved us less. My kids understood that and looked forward now to having her to themselves for a while.

She wanted to go home. There was no doubt about that. Andrea and I wanted that for her as well, but there just wasn't a way to work that out. She finally, reluctantly, agreed to come and live with my family until she could go home again. She longed to be back in the house she loved—her home. I understood that and . . . I wish now I would have figured out a way to make that happen for her.

Moving day arrived. My brothers and Andrea brought her belongings and walker into the house—into our TV room, which I had prepared for her to be as homey as possible. Richard had repaired and painted the stained ceiling, and we'd brought in a bed, I put her grandmother's rocking chair in there and made her bed with a pretty quilt my mother-in-law had made for Truman. I knew she liked to lie in bed and look at pictures of her grandchildren arrayed on a shelf nearby and pray for each of them every morning, so I moved the books off the bookcase and placed framed pictures of each grandchild where she could see them from the bed.

My brother Paul lugged her reclining chair in first and put it in the living room. Mom was helped to it and left there as we carried and arranged things. We were oblivious to one of our barn cats, Bill, who saw his chance in the propped-open door and came in, checked things out, and decided that Mom's lap was the warmest and softest place in the house. He was correct about that and, surprisingly, Mom didn't mind. She'd always said cats belonged outside, but here he was, soft and babylike, purring in her lap, and she smiled sheepishly at us, petting him with her twisted hands.

That's my favorite memory of those weeks—watching her hold Bill—that and the evening she was able to attend Marian's Christmas concert with us. She could walk, but it was a long way from the curb to the school door, so we set up the wheelchair and Truman zoomed her up the sidewalk, both of them laughing. Indoors, a few girls approached and invited him to sit with them. "I'm gonna sit with my grandma," he said—not unkindly, but as if he were really looking forward to the opportunity to do so. Mom hid her smile, glancing over at me to see if I'd noticed that small teen drama, purring a bit herself at having such a handsome grandson to squire her about.

At home, I'd taken up all the throw rugs in the house and made a path for Mom to do laps between her recliner and the piano, the window seat and the couch. She knew her return home depended on how strong she could make herself, and she powered through the circuit several times, over and over, each day. Then she'd sit down and Bill, who was now a delighted frequent visitor to the house, leapt into her lap.

Returning home was never to be, though. She became confused, weak, uncertain. I took her to the clinic, and when the doctor diagnosed depression, I remembered what Mom's friend Vernice had said about that: "Of course we're depressed," she growled in her old-smoker's voice. "We're OLD." To the doctor, I made the case, horrified that my voice shook and broke, that this was unusual behavior for Mom—that she had been doing very well, that she had SOMETHING going on that should be tested for and diagnosed.

He relented and admitted her to the hospital. Tests revealed hypercalcemia—her already porous bones were releasing calcium into her bloodstream, messing with her ability to think.

And that was the beginning of the end. It was the end of Mom living with us. She died in a nursing home in February of that year. My sister and I spent the night beside her. We sang hymns from memory, and Mom gazed up at us, unable to speak by this time except through those beautiful eyes.

When it was over, after Marian packed up her cello to bring it home after playing at the funeral, after Andrea and I had folded the quilt Mom's grandmother had made and which we'd used to cover her casket, after I'd gone home and the kids were back in school and friends began to greet me as they always had, rather than with concern for my recent loss—I began to fall apart inside. No one knew, least of all Richard. I lay in bed at night with tears leaking from the corners of my eyes and into the pillow as he slept. Had I awakened him, he'd have been concerned—of course! But then I'd have had to articulate what I was feeling, which was this: like a systemic failure. How do you explain that? And I knew what he'd say, what anyone would say, pretty much. "No, you're not!" And then a litany of my accomplishments, none of which were anything to write home about. As if home still existed.

***

Spring came, as it does, and with it all the work it always brings. Small wildflowers—bloodroot, hepatica, violets—bloomed. Andrea and I used to pick them for Mom; her birthday had been in April, and quite possibly the proprietors of the flower shop in our town knew it as a harbinger of spring, that Marian Ash's daughters would be in to buy a small vase in preparation for the wee, sweet bouquets we planned to gather for her.

On Earth Day, which was her birthday, I cleaned up one side of Cady Lane and down the other, hauling aluminum cans, beer bottles, and food wrappers behind me in a trash bag until it became too heavy. Then I left it to be picked up later and continued with another one I'd brought in my backpack. I did it, and still do it, in honor of my mother. And I laugh inside because picking up other people's trash, thrown selfishly out the window, always reminds me of something sarcastic she said once about litterbugs: "You wouldn't want them to have a dirty car!" And I find myself growing impatient and hateful toward other people, which is the opposite of how Mom was, and that reminds me of her, too.

A year later Marian played her cello with the community orchestra in the university's concert hall. Their final song was the most familiar part of Handel's Messiah, the *Hallelujah Chorus*, which had been my mother's favorite song, in part because it had been her dad's favorite.

The first chords brought all of us to our feet—except Richard, who was unaware that one stands to hear this music. Puzzled, he stood as well, and we watched our daughter, head bent, bowing that beautiful instrument among the other musicians. The warm wood of the concert hall, the music, thinking that if Mom could have lasted one more year she might have been there to see this, Marian's serious expression as her fingers moved to make the notes—my neck was wet with tears. I was still damp-faced and red-eyed when Marian met us after the concert. I didn't have to explain it. We just went home. I don't think the part of me that tore when she died has ever or will ever heal. I'm sure it is like this for everyone who has lost a parent—I'm nothing special. Well, that's part of why I grieve. I wish I'd been more for her.

Polar Ice

My mother came to live with us
so she could gain strength to live
on her own, that was the plan, go
home to her own house
after living in ours
after that stint in the hospital.
In the ruckus of moving her,
our barn cat Bill slipped
through the door and saw
his chance in her lap, Mom nicely
settled in the comfy chair as
we hauled in her stuff.

She always said cats belonged
outside but he was warm,
about the size of a baby, and he
just wanted to be near her, purr
under the bent twigs of her hands
and the rest of the time
she was with us Bill lived
in her lap.

Neither lasted the year, though
Bill made it to October before
he went missing. A coyote, we
figured as the winter passed
and on each trip through the harness room
he did not lie curled
on the feed box, showing sleepy interest
in our work.

We found his dry hide and bones
under a wagon in April, I buried
him by the house and planted
a rose bush over the grave.

It's a hardy rose, named Polar Ice,

the palest pink, as if—like this
hankie I have of my mother's, like
this memory of her finding solace
in an old barn cat—
the white petals had been washed once,
long ago, with something
vivid.

Chapter 23

I did not run for town supervisor again. That spring was my last meeting after eight years of being Supervisor II. At first I felt uneasy on the first and third Mondays of the month, as if I'd let someone down by forgetting something important. Then those evenings became just plain old evenings like any other.

I realized how much time I'd spent at that job when I received a check from the town clerk for meetings other than town meetings I'd attended. I was paid $119 per month for being a supervisor. At its most basic, that job entailed attending 24 town meetings per year. I put in several hours a week on it outside of the meetings, though. I'd taken courses in land-use planning at the local university, hoping to learn more that would help me be of greater service to my town. I attended Wisconsin Towns Association conferences and land-use conferences, went on my own to talk with experts in land use, and read everything I could on the subject. And I'd been heavily involved in preparing our town plan for the future—a lot of us worked on that. There were meetings and meetings. My check was for $750, which meant that at a rate of ten dollars per meeting, I had attended 75 meetings outside of the town board meetings that year. And there were some meetings I attended that were not included in that check.

As for what I'd accomplished in all those hours, who can say. We'd established a bar for building on Exclusive Ag land through those difficult first meetings. It never got so ugly again while I was there.

I was still on the county board, though, serving the second year of a two-year term. I hated it and knew I would not run for another. The board consisted of seventeen members, most of whom had been there for decades. There was a way to do things, and logic had no part in how they made their decisions. Land-use planning, for example, which was my main reason for being involved in local government, was based on the county plan, which was that all the towns—save mine, which had opted to administer its own zoning—were free to make planning and zoning decisions based on their own plans, which were the county plan. This circular system meant that no one could really know what was going to happen across the county in the future, as far as development or road usage.

I survived to the end of my term. Richard and Truman were away every non-winter weekend for mountain-bike races all around the state. Truman had balance and athletic ability far beyond anything we could have bequeathed to him genetically. It was astonishing to see him race the bike up and down hills most people would navigate on foot with difficulty.

He was becoming an accomplished bike mechanic, thanks to his early mechanical work on the farm and to his co-workers at the bike shop. He collected bicycles and worked on them at home and at the shop. He welded two bikes together—normally I didn't worry much about him getting hurt, but that one, with him riding down our steep driveway high in the air on that Rube Goldberg contraption, gave me some pause.

He and his friends Ryan and Bobby were busy all morning on some project on the sledding hill in the geldings' paddock southeast of the house. Later in the day they came into the mud room and called me down from where I'd been working in my upstairs office. Truman was sitting on the bench, head down, breathing deeply. Ryan and Bobby were alternately snickering and wide-eyed with concern.

"We made a ramp," they told me. "Truman crashed off it."

"I think I hurt my neck," Truman said.

"How high is this ramp?" I asked.

"Ummm," the boys pondered. "Ten feet, I think."

Well, Jesus H. Christ.

"I WISH we would've got that on video," I heard one of the boys say as Truman and I got into the car for our trip to the ER.

Well, he was only shaken up. The first x-ray showed a nick out of a bone in his neck, but further angles didn't support that conclusion. Dr. Goblirsch reassured me that my son was okay. Then he turned to Truman: "Hey, great job on that tune-up," he said. "My bike shifts perfectly now."

***

It was the wee hours of the morning, the usual time for a mare to be foaling on our farm. Margaret, though she'd had several foals already, seemed to be having problems with this one. She got up and lay down, got up and lay down, sweated herself wet.

I decided it was time to stop observing and start participating, however I could. I slipped into the foaling stall and put on a long plastic glove from the foaling kit. Margaret was lying down, panting.

"Hey, girl," I said. "Seems like a lot of work for you." I knelt and stroked her backside with my ungloved hand, then slipped the gloved one inside her.

Not too far in I could feel the foal's hooves.

"Got a baby in there," I said to Margaret, who was on her side, head in the straw. "I can feel it. Let's see..." and I did feel it, the hooves, except they weren't as I expected them to be. The foal was not poised like a diver to slip out of its mother. This foal was on its back. That was the problem.

Well, this wasn't the first time for that on this farm, starting years ago with the birth of Tug. I felt for the baby's nose, which was positioned to follow the hooves. Good. If the head were bent back, I'd have to get Dick out of bed to help me with this. I reached for those hooves again, but they were farther back inside Margaret.

"Margaret," I said. "I can help you out, but you need to push a bit so I can get a grip on your baby's legs, okay?"

She picked up her head and looked back at me. The next contraction moved those hooves to within my reach, and I could grab above the fetlocks, pulling one ahead of the other down toward her hocks, the foal turning as the next contraction moved it closer to being born, and then that big splat/plop as the shoulders came out and the foal was essentially born, flailing its head, dislodging the birth sac, looking like a wet baby moose somehow sticking out of a beautiful mare.

How very much I loved this—the beginning, the birth, the thing that happened next: Margaret's low whicker as she looked back and saw her baby and fell in love. The rest of the foal would be out soon, I'd treat the navel with a mild antiseptic, and I'd see that this was a filly we'd already planned to call Hannah Rose. She'd stand and find her mother's udder, and I can tell you, because I am writing this years later, that she went on to have a good life. Some births did not work out that way, and that healthy one filled me. I will never get over how much I love this farm.

Back in the house, I reported on the foal to Richard, who was headed up the hill to check on whether it was dry enough to plow.

I saw that it was time to wake the kids up for school. Truman was harder to rouse, so I started with him, stepping over scattered bike magazines, bike parts, assorted tools, and what I'd come to think of as Truman Stuff—little gadgets he'd assembled, such as the string and pulley and electric motor thing he made so he could turn off the overhead light from his bed.

"Bud, time to get up." I tugged at his pillow gently at first and then harder, until I saw that he heard me and knew it was time to wake up.

Then on to Marian's room, which was also cluttered, but in this case with collages on the walls—mostly of animals—clothes on the floor, and books piled beside the bed. She awakened more easily because, as it turned out, she had an agenda.

"Mom," she said sleepily. "I have to meet Steph early to practice for our ensemble contest song."

To be honest, I wasn't really listening. I was thinking about that foal in the barn. "Margaret had her foal last night," I said. "A filly."

Marian was up on her feet now, hands in her hair, waking up. "Oh, good," she said. "So I need a ride to school early today."

Wait. I was still not really paying attention. "Today?" I asked, trying to orient myself to what was happening there in the house, not out in the barn.

Marian sighed heavily. Her eyes rolled in a way with which I had become very familiar now that she was sixteen years old. "Yes, Mom," she said, slowly and clearly. "TODAY. That's what 'today' MEANS in Marian language."

I couldn't help laughing, but I tamped it down. On my way to the stairs I looked through the open door into Truman's room—he was sitting up, groggily swiping his armpits with a stick of deodorant. First things first. Where did he keep that stuff? Under his pillow?

And so another day started for us on Baldur Farm.

Chapter 24

One day I was standing in the kitchen when Richard came in to tell me something. Was it raining? It would be almost too poetic, wouldn't it, if it had been? He said, "I'm done growing potatoes. I'm sick of fixing broken equipment. I don't want to do it anymore."

And I said okay. He's almost eleven years older than I am. I was tired myself.

In March of that year he did not drive to Colorado for organic potato seed. We began selling horses, starting with my beloved team of geldings, Aldo and Jed. Luckily I found the best possible home for them and have remained in touch with Bill, their owner. They've had the best of care and will have it through all their lives. Someday Bill will call to tell me Aldo has died. I know what I will do then—walk up into the old geldings' paddock and sing, low and slow, the way I used to when I went up there to bring him and Jed down to work—Bridge over Troubled Water.

We all knew Richard would find a new passion, but he surprised us with what it was—making flutes in the style of those made by the indigenous people of this continent, mostly in the Hopi style. They are beautiful but hard to play—just thin tubes of wood he has cut and hollowed out, with a few holes for the notes. He can't stop making them, and I tease him that he has now made more flutes than there are flute players in the world who can play that style of flute. This bothers him not at all. He's working on another one right now as I write.

After my mother died, I took solace in baking bread. It became something I had to do, and as I experimented more I learned about brick ovens, and I decided to try building one. Richard helped me pour the concrete base and then (at my request) left me alone with a pile of bricks. It took me weeks to build it, and it has had to be remodeled twice since then due to my poor workmanship. But I've learned to fire it and make dozens of loaves in a day with it, with flour I grind myself, often from grain grown on our farm. Which brings us back to the garden at our house on the lake, the patch of wheat, my dad sweating through his good hat as he showed us how to make sheaves and shock them to dry.

We're on sabbatical from the farm at the moment, staying in a cabin in southern Oregon. One of the first things I did here was mix flour and water and leave it to take in yeast from the air. It took a few days, but I now have a culture. Just an hour ago I fed it and put it in a sunny part of the kitchen. I forgot about it and then, a moment ago, I looked at it and found it overflowing the jar, active and wild. Kind of like my heart when I hear the horses running up from the lower pasture at home, or here, when I walk down to the river and watch it smash into froth on the rocks.

***

I seldom have epiphanies. Surely in writing all this down I should have had one or two? And what was the purpose of the writing? I told Marian I was going to do it, and she said, "Well, we'll read it." And maybe that's enough.

Or maybe my epiphany happened years ago. Sometimes, back then, at night before I fell asleep, I thought about what could have happened that day, had any of us been badly hurt or killed, or the house had burned, or we'd caused terrible harm to someone else; I thought of how desperately I would want just this—the darkness, Richard's belly against my back, both kids in their beds sleeping safely, deeply—just this. What I had.

On solstice last year, Richard and I walked up the farm road after nine o'clock in the evening. It was light enough to see—even to photograph—the deer we saw crossing ahead of us with her fawn. The light seemed elegiac, and maybe that is why I have chosen this memory as the one that will close this book. Climbing the slow rise of the farm road, we passed, in what used to be the geldings' paddock, newly turned ground that Richard wants to plant as a small hayfield. The top of the rise is where both our kids were married, in a grove of box elders down which, for some reason years ago, Richard mowed an aisle, and the remaining trees grew in pairs, linked at the base, their leaves creating a cool, green, light-filtering canopy. Farther along we looked out at the south forty, now planted to hay, corn, and a cover crop. Truman is farming this place now, in addition to his work as a technical expert at a large bike-tool company—a job we now know, looking back, he was preparing for since he was eight, studying those bike-parts catalogs. He uses tractors, not horses, and I understand. "You and Dad did everything the hardest way possible," he explained to me once, and I can't really argue with that.

Marian studied biology and parlayed that into a series of outdoor, animal-related jobs that culminated in training dolphins for the US Navy. Now she lives in Italy, teaching English as her husband completes his PhD, after which she will, I am sure, find a way to work with animals again.

That solstice walk brought us back down past the pasture, smaller now that we have only four horses. I have kept my promise to Katrina. Patience has never left the farm, and never will. Still beautiful, she stands with Damon most days. They are willing to work when asked (which I do sparingly, given our ages and the tasks left to do on this now motorized farm). A few years ago I bought a mare named Patches, because she had gotten her previous owner through a rough patch, and I bred her to a Suffolk stallion named Sweet William. She had a colt, coming two this year, Harvey, who shows signs of being a bit like Martin—which should worry me a bit but instead just endears him to me. I'll buy a mate for him in a few months and train them to work. I've staked out a piece of ground. I have some ideas I still want to try.

And I have a grandson, Truman and Ashley's boy, who will need to see horses working, who will learn to pick up spuds, into whose small hands I will one day fit the worn leather straps. Not with designs on his future, not willing him to follow our lead—but merely to teach him to hold the lines.

***

Farm Truck

Well, we're spending a rainy morning in the truck.
Mr. and Mrs. Old Farm Couple
driving to buy seed and to visit my niece,
bundling the trips to save on fuel.

Spring and a light rain,
corn poking up like rows of lit matches
burning green out of the ground, today's drizzle
like gasoline to their little flames,
just weeks till this countryside is torched
with corn.

Windshield wipers flop and rest,
flop and rest. One work glove curls on the floor
fingers flexed as if it still wants to grip
the pliers or twine. Tool box, a tarp
to cover the seed once we've bought it,
a railroad spike we use sometimes
when hauling a hay rack.

On the freeway we putz
in the slow lane, passed in flumes of spray
by cars and tightly muscled newer pickups
almost perky in their powerful lines.

The long crack in our windshield shines
silver as our hair. Not far now.

Published by Acres U.S.A. Books

Acres U.S.A. is North America's oldest publisher on production-scale organic and sustainable farming. For more than fifty years, our mission has been to help farmers, ranchers and market gardeners grow food profitably, regeneratively and without harmful chemicals.

We offer a full range of educational products—including books and pamphlets, a monthly magazine, webinars, podcasts, online courses and both large-venue and on-farm events—to help producers find timely, practical and well-tested knowledge and advice on just about any agricultural topic.

Acres U.S.A. is committed to providing information about agricultural production systems that can regenerate the soil, support the farmer and rancher, and bring healthy, nutritious food to the consumer. We bring both new and time-tested voices to the discussion of "what is to be done" when it comes to building soil health, raising healthy animals and growing nutritious crops. We amplify the work of modern agricultural leaders and innovators, and we support all efforts to increase the resilience of our farms and ranches—and the people who anchor and feed our nation.

The books we sell, and the speakers at our events, and the editors of our monthly eNewsletter and farmer-facing website represent generations of experience in building—not mining—soils. The lessons they teach are fundamental: to farm economically, one must do so ecologically.

Visit acresusa.com
to learn about our magazine, events, books, and other educational opportunities.

Other titles from Acres U.S.A. books:

The Biological Farmer
By Gary Zimmer

Biological farmers work with nature, feeding soil life, balancing soil minerals and tilling soils with a purpose. The methods they apply involve a unique system of beliefs, observations and guidelines that result in increased production and profit. This practical how-to guide explains their methods and will help you make farming profitable and fun. Biological farming does not mean less production; it means eliminating obstacles to healthy, efficient production. Once the chemical, physical and biological properties of the soil are in balance, you can expect optimal outputs, even in bad years. Biological farming improves the environment, reduces erosion, reduces disease and insect problems, and alters weed pressure and it accomplishes this by working in harmony with nature.

Natural Horse Care
By Pat Coleby

Proper horse care begins with good nutrition practices. Chances are, if a horse needs medical attention, the causes can be traced to poor feeding practices, nutrient-deficient feed, bad farming and, ultimately, imbalanced, demineralized soil. Pat Coleby shares decades of experience working with a variety of horses. She explains how conventional farming and husbandry practices compromise livestock health, resulting in problems that standard veterinary techniques can't properly address. Natural Horse Care addresses a broad spectrum of comprehensive health care, detailing dozens of horse ailments, discussing their origins, and offering proven, natural treatments.

Weeds—Control Without Poisons
By Charles Walters

"Low biological activity is inherent in each weed problem ... Each weed is keyed to a specific environment slotted for its proliferation." So says Weeds—Control Without Poisons author Charles Walters. Further, calcium, magnesium, potassium and other elements in equilibrium are likely to roll back more weeds than all the available herbicides on the market.

Specifics on a hundred weeds, why they grow, what soil conditions spur them on or stop them, what they say about your soil, and how to control them without the obscene presence of poisons. All cross-referenced by scientific and various common names, and a pictorial glossary.

Small Farms Are Real Farms
By John Ikerd

Since the middle of the last century, American farm policy has taken the nation into the dead end of industrial farm production and food distribution. Farming, at its core a biological process, has been transformed into an industrial process, thus demolishing the economic and cultural values upon which the nation was founded.

Along the way, small farms have been ridiculed and dismissed as inconsequential—but now the seeds of a rural renaissance are being planted, not by these industrial behemoths, but by family-scale farms. In this collection of essays by one of America's most eloquent and influential proponents of sustainable agriculture, the multifaceted case for small farms is built using logic and facts.

www.ingramcontent.com/pod-product-compliance
Lightning Source LLC
Chambersburg PA
CBHW071943160426
43198CB00011B/1525